THE BATTERY-POWERED HOME

THE BATTERY-POWERED HOME

FOOLPROOF GRID-TIED LITHIUM STORAGE

GREG SMITH

HOUNDSTOOTH
PRESS

THE BATTERY-POWERED HOME

Foolproof Grid-Tied Lithium Storage

ISBN   978-1-5445-2159-6 *Hardcover*
         978-1-5445-2158-9 *Paperback*
         978-1-5445-2157-2 *Ebook*

For my loving wife, who has put up with me all these years, in and out of the military, and gave me the courage to finish this book.

To the men and women in the renewable energy industry. You are making the world a better place.

CONTENTS

A true teacher would never tell you what to do. But he would give you the knowledge with which you could decide what would be best for you to do.

—CHRISTOPHER PIKE, *SATI*

DISCLAIMER

The views and opinions expressed in this book are those of the author. They do not necessarily reflect the views or opinions of present or past employers, media outlets, organizations, companies, websites, or other sources mentioned or cited in this work.

Although the author and publisher have made every effort to ensure that the information in this book was correct at press time and publication, they assume no responsibility for errors, inaccuracies (including sea stories), or omissions and hereby disclaim any liability for any loss, damage, or disruption caused by errors or omissions, regardless of cause.

References to brands, products, or services do not, explicitly or otherwise, imply endorsement or recommendation by the author. Brands, products, and services are mentioned as educational references only. Manufacturers did not pay to have their names or products printed in this work. Therefore, the author is indemnified of any hurt feelings, judgments, or pouty behavior.

This work may contain content not authorized by its owner. The author and his legal representation declare fair use of the infor-

CHAPTER 1

INTRODUCTION

Robert and Carol Beckett were sitting with their two young children in the dark. Through the living room window drapes, they could see the culprit of their power outage—the faint orange glow of a wildfire only 20 miles away. The winds were in their favor for the time being, and they were safe. It had been another 100-degree day in the Sacramento Valley, and without air conditioning, the house had reached an uncomfortable 90 degrees. But they dared not open the windows since the soot and smell of the wildfire would quickly enter the house.

Robert used his cell phone sparingly, turning it off when not in use, but Carol kept hers on for fire alerts.

The faucets still worked for now. Carol quickly opened the refrigerator only when necessary. She finally found the old handheld can opener since the electric one was useless. Robert lit candles and used his old camping lantern when walking around the house. There were a thousand different ways the Becketts would have liked to have begun their weekend, and this was not one of them. Then, an alert came through on Carol's phone.

The power outage was estimated to last another four days.

"How can this be?" Robert lamented. "We spent over $20,000 on these batteries, and here we are in the dark. I just don't get it. That salesperson said we would be able to make it through a power outage, but we didn't even make it a day." Robert had a severe case of buyer's remorse. He would call the installer first thing Monday morning since he couldn't get anyone on the phone over the weekend.

"Pack it up. Grab all the food you can. We'll find a motel."

As the Becketts drove off, Robert looked at his house with resentment. He saw the night sky's orange reflect off their rooftop solar array. "Piece of shit."

Although the utility power had been out for about 24 hours, the Becketts had had power in their home until eight hours ago. Someone had sold them a battery system that was supposed to provide power during an outage so their lives could go on with minimal impact. Which they did until the Becketts prematurely emptied their battery. It wasn't the battery's fault. It wasn't the solar array's fault. In the end, it was the Becketts' fault. But the solar company that sold the Becketts on disaster resilience had some responsibility to bear as well.

Unfortunately, this is an all-too-familiar story to me. If I had a beer for every email, phone call, or text I received from someone with this story, I would be a drunkard. In my 12-plus years of training solar and storage professionals, the dangers of lousy customer-expectation management have always been a priority. Although a technical trainer by job title, I frequently take service calls from frustrated installers because I am usually the first technical person they meet. Occasionally, I am put on speakerphone so the homeowners can also voice their frustrations.

I have been a trainer for most of my adult life, beginning with my first naval shore-duty assignment as a submarine sonar instructor. In fact, I spent every shore duty I could as an instructor since I had so much fun with my first assignment. I had to learn how to train and speak using language baby sonarmen, senior enlisted, and officers could understand since they would usually all be in the same room together. I have always maintained that anyone could learn the material if I could make my training sailor-proof. If you can get a sailor to understand and safely execute a new skill without hurting himself or the equipment, the curriculum is solid. That is how I wrote this book.

This book started as a LinkedIn blog post, but I realized it could be much more than that. I want to help you be a better solar and storage professional. I want to emphasize the word "professional." This book is *not* for do-it-yourselfers who want to punch holes in their roofs for cheap Chinese solar modules to charge golf-cart batteries with a $30 charge controller to power their off-grid shanties. I've seen too many of those dumpster fires to know better and will leave that part of the industry to someone with more guts.

This book is *not* for solar and storage system installers—the people on the roof who fasten solar modules, bend conduits, turn wrenches, and run wire. Those skills require hands-on on-the-job training (OJT) and are beyond the scope of this book. If you want that type of training, I highly recommend attending a Solar Energy International workshop.[1]

This book is for the professional who has been cautious about adding storage to their current solar portfolio. It is also for the professional who has had a bad experience installing storage, like unhappy customers or poor product operation. It is for the person who is having difficulty navigating the swift currents of storage in general.

Prepare to be schooled beyond the spec sheets, beyond the sales and marketing hype. Prepare to talk to your customers in an intelligent and informed way that establishes and maintains your status as a subject-matter expert. If you find yourself saying, "That's a great question. Let me get back to you on that," more often than you are comfortable with, keep reading. I get weekly calls, emails, and texts from installers about their unhappy customers. Poor customer-expectation management is usually the culprit. Bad system design is another offender. I will cover these pitfalls in great detail in this book.

If you size and design solar and storage systems, I have devoted an entire chapter to your craft. Mainly because this is where many negative customer-experience issues can be avoided, particularly with battery systems. I will show you how to size these systems to keep you, the customer, and the product manufacturer's service line happy.

I will also put my sales cap on as I explore the following presale questions homeowners ask most often:

1. How much does it cost?
2. How long will it last?
3. How long can I power my house?
4. How does it work?

Solar is one of the fastest-growing industries worldwide, and storage is not far behind. There are many places in the United States

(Hawaii, for example) where installing solar without batteries doesn't make sense. In 2019, only 5% of installed solar systems were paired with storage. By 2025, that number will grow to 25% (according to Wood McKenzie).[2]

The reasons to dive into this industry are as numerous as the products and jobs it creates (250,000 American jobs at the time of publishing). As climate awareness grows, so does the desire to reduce our carbon footprint. Solar and storage offer a clean and reliable energy source that appeals to the climate sensitive and environmentally responsible—and having a miniature power-production plant in one's home appeals to the techno-geek in all of us. Combining solar and storage offers true grid resilience, peace of mind, zombie-apocalypse sustainability, and energy security. Among all these legitimate reasons is the real opportunity to make a good living in this billion-dollar industry.

The internet has closed the gap between novice and expert; however, beneficial information is intertwined and buried within the sales and marketing fog or the court of public opinion. Navigating these waters can be a daunting task, and this book is the nautical chart to keep you from running aground.

Even in the renewables industry, storage is still relatively new to many people and downright scary to some. This apprehension seems counterintuitive since solar and storage started together decades ago in the quest to save the world from fossil fuels. Or maybe it was to hide grow lights? Regardless, some solar companies deliberately shy away from storage (offering, promoting, or even talking about it) until they get a pesky customer who wants it. Why would someone intentionally turn away business? For them, storage is too much of a hassle, there isn't enough margin, or they lack the skillset to size, design, and install the system. It becomes a scramble to learn storage "real quick" when they do get a serious customer.

Newsflash: those customers are becoming more frequent!

You must educate yourself and your team on storage—there will be a time when installing solar without storage in your area won't make sense. You should be prepared for this inevitability even if you do not think it is imminent. You do not want an angry Robert Beckett chewing you out.

"Cripes, man, just get to it already!"

Yeah, that's him. Choose your own actor, TV personality, or animated show character for his voice. I have mine, but my lawyer advised against mentioning it. Let's dip our toes into the murky waters of storage and start making sense of it all.

CHAPTER 2

TRONS

Regardless of your industry experience, we need to speak the same language. Let's start with some definitions and concepts that will provide context for the rest of this book.

"You are going to start with definitions? What are we, grade-schoolers?"

I thought about skipping this, but after seeing industry professionals confuse kilowatt (kW) with kilowatt-hour (kWh), I decided this is an excellent place to start. This confusion is akin to asking how fast is 563 miles (the distance from my house to Las Vegas, baby!). A glossary shoved in the back of this book would probably not get the attention it deserves.

INTRODUCTION TO ELECTRICITY

Two types of electricity are generated, converted, and stored in solar and storage systems. While each has unique properties, the common denominator is they can kill you and probably hurt you the whole time they are killing you. Most people don't have any idea what a volt is, know what a watt comes from, or understand what their electric company charges them for each month. But we all know how much we pay! These are all critical concepts to understand; otherwise, you will be in the dark for the rest of this book, but more importantly, you will be lost when trying to talk to someone about solar and storage.

THE BASICS

Three unseen forces move electrons (electricity) through a wire and into your 65-inch flatscreen so you can watch your favorite streaming service. Water flowing through a pipe is a universally accepted analogy for electricity movement that transcends all races, ethnicities, religious preferences, and favorite *Rick and Morty* episodes.

The trustworthiness of electron behavior is best illustrated by a quote from my sea dad back in 1989 aboard my first submarine, the USS *Los Angeles* (SSN 688). He talked to me about circuits and sonar fault localization and said, "Trons are trons, Smitty. They have to follow the same rules no matter where they are."

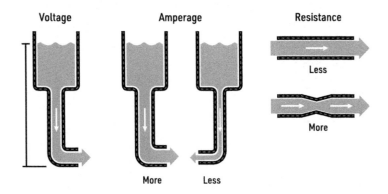

RESISTANCE

The resistance to electron movement, expressed in ohms, is called, well, resistance. Yay, for self-defining terms! Rubber is an excellent insulator, which is why electrical safety equipment uses so much of it. A rubber glove resists the flow of electricity from an energized breaker box to your bare hand. The only part of an appliance power cord with exposed metal are the prongs. The rest of the power cord is covered in a mixture of insulating materials to keep you and the equipment safe. The diameter of a pipe represents the resistance in an electrical circuit. The larger the diameter, the more water can flow, the lower the resistance. Keep this in mind when we start talking about wire sizes.

VOLTAGE

Voltage, expressed in volts (V), is the force that moves or pushes electrons, analogous to the water pressure in a pipe. For example, your house's water pressure is much lower than the water pressure in the nearby fire hydrant. The two typical voltages in your home reflect these two varying water pressures. We use 120 V outlets for small appliances and device chargers and 240 V outlets for larger loads, like an electric dryer or central air conditioner. Ever wonder why the electric dryer has that big, oversized plug, while everything else uses a two- or three-pronged plug? Mainly so you won't plug your phone into a 240 V receptacle and blow it up. More importantly, it is because the dryer heating element needs a higher voltage, or more push, to dry your clothes than required by smaller appliances.

CURRENT

The rate at which the electrons move, expressed in amperes (amps, or A), is called current. It represents the flow of water in our analogy. Larger loads need more amps, which means there is more water flowing in the pipe. Current is what does all the work in an electrical circuit. It causes the light bulb to glow, the air compressor to start compressing, and the mobile device to come to life. Everything in an electrical circuit is rated to carry a specific amount of current pushed by a rated voltage. Devices that use more current must also use larger wire sizes.

Some materials are better electrical conductors than others, meaning some substances allow current to flow through them while reducing the heat that is the natural by-product of the flow of electricity. This buildup of heat is called friction, and I bet you have felt an appliance plug or the plug on your mobile device heat up at one time or another. Copper is the best conductor of electricity, so it is the predominant material used in most circuits. However, copper fetches a pretty penny, either from the electricians buying it or for the criminals ripping it out of unpatrolled construction sites. Aluminum is used for large wires to save money, and they are double the size of their copper-wire equivalents. Steel is also used, but mainly in utility transmission wiring since they use so much of the stuff. If they used copper, our utility bills would be ridiculously higher.

Suppose you exceed the current rating of a device. Current flow suddenly becomes visible as the wires catch fire. As homeowners, we do not have to worry about these ratings since manufacturers build products to the correct specifications. Solar and storage designers must do some quick math to ensure they are using the right-sized wire and breakers to protect the circuit.

AC AND DC

Two types of current exist in the known universe—alternating current (AC) and direct current (DC). Even if you did not know this, the acronyms should, at the very least, trigger images of a large bell being struck followed by a low, ominous ringing. (RIP Bon Scott.)

Generated by the utility company, wind turbines, water turbines, or generators, alternating current is the type of electricity that powers your home. It gets its name from the behavior of the electrons in the wires they travel through. Alternating current moves back and forth, changing direction at a specific rate or frequency. The grid frequency for most of the world is 50 Hz (hertz), although the United States and a few other countries use the 60 Hz standard. In our industry, solar inverters make alternating current. More on these later.

Direct current acts quite a bit differently than its AC cousin. It only moves in one direction and has a frequency of zero. DC is more commonly used as stored energy in batteries and is typically not produced or transmitted by a utility company. Solar modules make direct current, as do other forms of renewable energy.

AC and DC have their pros and cons. Ultimately, Nikola Tesla's AC beat Thomas Edison's DC patents. Alternating current became the de facto power source for most of our planet mainly because AC proved to be far superior in power transmission over long distances.

OHM'S LAW

There is such consistent and measurable behavior between the forces of current, voltage, and resistance that German physicist and mathematician Georg Ohm created a handy formula to express their relationship back in the early 1800s.

I = E / R

"I" represents current, "E" represents voltage (don't ask), and "R" represents resistance. If you have two of the three values, you can always find the third. Yay, algebra!

POWER

Simply put, power refers to the number of watts (W) being generated or consumed. A 5,000 W generator from your local hardware store indicates it can provide 5,000 W of continuous power. A microwave consumes 1,000 W to heat your frozen Hot Pocket. Therefore, that generator could power five microwaves for a room full of gamers pulling a *Call of Duty* all-nighter.

In 1841, James Prescott Joule reworked Ohm's law to create the formula for electric power. If Joule's name sounds familiar, you are probably thinking about the 21 gigajoules (GJ) required for time travel. Electrical power (P), expressed in watts (W), is the product of voltage (V) and current (A). This formula is used a lot in this industry, and, like Ohm's formula, you can always find the third value as long as you have the other two. This graphic visualizes all three possible formulas.

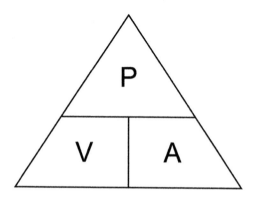

I (mistakenly) used to call it the Power Triangle, but an electrical engineer quickly put me in my place. And he was right to do so. The actual Power Triangle involves a much deeper level of electrical theory and includes real, apparent, and reactive power. I could tweak the shape and call it the Power Reuleaux Triangle, but I don't think it would catch on.

Shifting the formula around ($A = W/V$), we can solve for amperage. For example, a 50 W stand-up-lamp bulb connected to the 120 V wall socket will require 0.41 A to operate.

Wattage often exceeds numbers that can be written out or even read comfortably, so we lean on the metric system to get rid of a few zeros.

Kilowatt—a thousand watts (1,000 W = 1 kW), the power requirement of a toaster oven

Megawatt—a million watts (1,000,000 W = 1 MW), the approximate power requirement of 150,000 homes

Gigawatt—a billion watts (1,000,000,000 W = 1 GW), one-twelfth the power requirement of a space-shuttle liftoff

"Dude. Space shuttle? A billion watts?"

As with most large numbers, it becomes a challenge to put wattage into perspective or define it using a relatable example. Unless you are in the business of commercial- or utility-scale solar and storage, kilowatts and kilowatt-hours are good enough. But we need to know how to find these values. For example, stamped on stickers on solar and battery inverters, their continuous power ratings indicate how much power they produce. Your appliances, lights, and other electrical devices require a certain amount of power. At the very least, knowing how much power you make and how much you need will allow you to figure out if you have enough to run the things in your house.

ENERGY

Energy, expressed in watt-hours (Wh), is the number of watts used over time (W times hours). For example, if you leave a 50 W light bulb on for one hour, it will consume 50 Wh of energy. If left on all day as a porch light, it would consume:

50 W × 24 hours = 1,200 Wh (1.2 kWh) per day

Let's investigate how much this bulb would cost in electricity if you left it on all day. Hawaii (the nation's highest electricity rate) and Louisiana (lowest rate) will be our two extremes.[1] Hawaii and Louisiana charge 32.76 and 9.37 cents per kWh, respectively.

Hawaii: 1.2 kWh × 32.76 = 39.31 cents a day, about $12 a month or $141 a year.

Louisiana: 1.2 kWh × 9.37 = 11.24 cents per day, about $3 a month or $40 a year.

That's quite a difference for one little light bulb. Maybe my grandfather was right to yell at me for leaving the lights on all the time.

Energy and power are often incorrectly used interchangeably. I have heard or read many people say something like, "I just installed a *40 kW* battery system for a guy living in a 5,200-square-foot home!" After asking some questions, I find out it was a *40 kWh* system. Huge difference. I can't imagine too many "average-sized" homes require 40 kW of power to run loads at any given time. But it is not uncommon for an energy hog of a home to consume 40 kWh of battery capacity in 24 hours.

THE RELATIONSHIP BETWEEN POWER AND ENERGY

Your utility company charges you for the generation, transmission, distribution, and, ultimately, consumption of electricity. You are also charged for the maintenance and upkeep of over 200,000 miles of high-voltage transmission lines, 5.5 million miles of local distribution lines, and all the necessary power-generation equipment. The utility meter keeps a watchful eye on how much energy you are using.

Utilities usually do not charge homeowners for the amount of *power* (watts) the air conditioner needs to run; they charge for *how long* (watt-hours) the air conditioner runs. But some utilities are shifting to a residential-demand charge, reflecting how much power the home will pull from the grid during a predetermined interval, usually 15 minutes. There are tiered rates for this power draw. The bigger the air conditioner, the higher the power surge, the higher the charge. And you are still charged for the time the air conditioner runs! Traditionally, only commercial buildings were subject to demand charges. The bad news is solar cannot hide these power spikes.

I recall a student in one of my commercial solar classes who installed 150 kW of solar on a warehouse rooftop. After the install, the first utility bill was a shocker because the owner was still paying $4,000 in demand charges. Unfortunately, this is not an uncommon story. These spikes in electrical usage usually occur in the morning as people come to work and turn on large loads such as air conditioning, pumps, and motors. The sun is probably not up yet, which means the solar system isn't producing any power to offset these massive loads. The spike happens, the utility meter records it, and boom goes the dynamite—a $4,000 bill.

"Dude, you're killing me with all this electricity stuff. I want to be a solar geek, not an electrical engineer."

OK, so let's talk about solar!

CHAPTER 3

PHOTOVOLTAICS

The photoelectric effect is an important conversation piece in physics, photochemistry, and electrochemistry. Oh, and it is also how a solar module works. Curiously, an internet search for "Who discovered the photoelectric effect" reveals a short list of scientists taking credit. Regardless of which nineteenth-century physicist discovered the photoelectric effect, the cell was woefully inefficient. Thus began the crusade for higher cell efficiency that continues to this day.

Solar cells are made of varying layers of semiconducting material (material that allows for natural electrical flow). Picture a big stack of buttermilk pancakes about the thickness of a human hair. The most common semiconductive material is silicon. Scientists continually experiment with various types of materials and the number of layers to boost cell efficiency.

About 93 million miles away from Earth, there is a generator with an inexhaustible energy supply and a 4.5-billion-year warranty. In the center of our sun, hydrogen molecules combine to make helium. The sun's core temperature and pressure are so high they fuse hydrogen molecules like stomping on a two-stack of different types of delicious ground meat. A by-product of this never-ending stomping is photons, the basic building block of light. Although it may take hundreds of thousands of years for these photons to finally escape the sun, it only takes them eight and a half minutes to reach Earth and slam into a solar cell to create electricity.

Fun fact: the sun's core temperature is 27 million degrees Fahrenheit (15 million degrees Celsius). That's the equivalent of 67,500 gas grills cooking metric tons of delicious brisket.

SOLAR MODULES

Dozens of solar cells are soldered together and placed behind tough, protective glass to make a solar module. *Many* people call these pretty blue rectangles solar "panels."

Photo by Michael Wilson on Unsplash

A solar panel heats water using copper tubes hidden behind the front-panel glass, but not all solar panels have that glass. Panels look suspiciously like a solar module from a distance, but a closer inspection reveals no individual solar cells. The closest parallel to the panel/module faux pas I can think of are people who use "barbecuing" and "grilling" interchangeably.

Here is a beautiful array consisting of 42 solar modules and 1 solar panel. Can you spot the panel?

But it's hard to take this seriously when module manufacturers use "panels" to describe their products.

"We manufacture the best solar panels on the market!"
— Every. Module. Manufacturer. Ever.

I've always wondered if this misstatement offends actual solar-panel manufacturers. To summarize: panels heat water, modules create electricity. Let's take this on-ramp and get back on topic.

There are many sizes and shapes of solar modules, as well as varying power outputs. While most modules are rectangle shaped, some manufacturers take aesthetics to heart and create flowers or circles. Or hearts. Higher module ratings equal fewer of them on the roof and more power in a smaller or similarly sized space. For example, I have sixteen 320 W modules on my roof for a total of 5,120 W of available array power. If I had sixteen 400 W modules, I would have 6,400 W in the same square footage. That is a 25% increase in power! But I'm getting ahead of myself.

pumps brakes

The following definitions allow us to understand further the effects of current, voltage, and power regarding a single module, a string of modules, and, ultimately, a photovoltaic (PV) array. Most of these values are used in overcurrent-protection formulas and expected-performance modeling. The first place to start is the specifications (spec) sheet of a PV module since this is where you will probably see these terms first.

Watts (STC)	300 W
Watts (PTC)	218 W
Max Power Voltage (Vmpp)	36.1 V
Max Power Current (Impp)	8.3 A
Open Circuit Voltage (Voc)	44.6 V
Short Circuit Current (Isc)	8.87 A
Max System Voltage	DC 600 V

Isc—Short-circuit current. The maximum amount of current the module can produce in a short-circuit condition. In theory, it is the equivalent of touching the positive and negative terminals of your car battery simultaneously with a crescent wrench, resulting in an impromptu class in arc welding.

Voc—Open-circuit voltage. The maximum amount of voltage the module is capable of producing. It is the voltage measured when there is no load on the module or array.

Vmpp—Maximum-power voltage. The point on the IV curve where voltage gives the maximum amount of available power under load.

Impp—Maximum-power current. The point on the IV curve where the current provides the maximum amount of power under load.

Let's start with the basics, which will explain why you see so many modules on a single rooftop. Let's say the module name-plate indicates a power rating of 300 W (standard test conditions [STC]). This means on a day with perfect conditions recreated in a lab (using industry-accepted STC methodology[1]), the module will make 300 W. If the temperature, sunlight, or thickness of the atmosphere relative to the solar module are different than the testing conditions, the module will not produce 300 W. To overcome the module's varying output levels and increase the power to something more meaningful, the designer will calculate the appropriate number of modules.

The Photovoltaics for Utility Scale Applications Test Conditions (PTC) rating offers a lower, more realistic performance value for solar modules than the STC rating.[2] Comparing the STC and PTC variables reveals some minor differences in the overall yearly production of solar systems. Systems in the wild rarely perform as the models predict, but people like the PTC ratings because they act as a cover-your-ass (CYA) model for managing harvest expectations. There are exceptions, of course, but I will let the banks, developers, and integrators hash that out with their engineers.

An array can be as little as 10 modules for a 3,000 W residential installation or 460 modules for a small commercial 146 kW installation to as many as 8 million modules covering 4,700 acres at the 550 MW Topaz Solar Farm in San Luis Obispo County, California.[3] Here is a shot of a solar farm somewhere between Sacramento and Phoenix that I happened to see from my window seat. A Google Map expert can find dozens of these farms.

Regardless of size, all that pretty blue glass is just a giant paperweight unless we can convert sunlight into electricity. This conversion is the primary job of the solar inverter, sometimes referred to as a "PV inverter."

SOLAR INVERTER TOPOLOGIES

Three primary solar inverter topologies convert the solar module DC into AC. They are string inverters, power optimizers (a string inverter used with a DC to DC device), and microinverters. There are significant differences between the three that are worth exploring.

STRING INVERTERS

String inverters get their name from the simple act of "stringing" modules together to produce power. Adding more modules increases the string's overall voltage, like dropping D-cell batteries into a flashlight. The negative conductor of one module connects to the positive conductor of its neighbor. The current of that string will not change regardless of how many modules are in the string. The voltage calculations use the open-circuit voltage (Voc) spec. Overcurrent protection calculations use short-circuit current (Isc). If you remember back to the power formula, if the voltage increases, so will the power.

But there is a hard limit to the number of modules you can connect in a string. It isn't so much about the number of modules but rather the overall voltage they produce when connected; 600 V is the limit for residential installations. This limit is dictated by the National Electrical Code (NEC) and enforced by electrical inspectors, local planning and permitting offices, and the inverter's capacitors. Exceeding the 600 V limit could cause the capacitors to explode like a soda can left in the freezer. The manufacturer's warranty does not cover this type of damage. Our sample module would allow 13 modules in series, but this is another rough-sized estimate since we have not compensated for any external variables.

If one string of modules will not allow us to reach the appropriate power level for our PV inverter, we will add more strings. Each subsequent string connects to its terminal block inside the inverter. Since we are adding strings in parallel, we can increase the current without going over our 600 V DC limit. My editing software tells me that I am using the word "string" too much, including this time, but I stand by it.

This process is called string sizing, and it is an age-old and time-honored practice in the industry. String sizing can get tricky, and there is a lot of math involved. Over the years, we have transitioned from manually calculating array sizes using pencil and paper to using elaborate spreadsheets or manufacturer software tools.

There are three primary applications for all PV inverters regardless of topology—residential, commercial, and utility scale. For string-inverter residential applications, ground-level mounting is the standard. For commercial rooftop installations, the string inverters can be on the roof next to the array. Utility-scale solar uses central inverters, so named because they receive the hundreds of strings that comprise the array.

A typical residential string inverter may weigh up to 130 pounds with a power output rating of up to approximately 10,000 W. Commercial three-phase string inverters get beefy. These inverters can reach power outputs of 50 kW or more. Central inverters can weigh over a ton and look like industrial equipment connected to thousands of solar modules. These gargantuan inverters can reach outputs of 2 MW, enough to power 400 homes.

MODULE-LEVEL ELECTRONICS

Microinverters and power optimizers form a unique group of solar products called module-level power electronics (MLPEs). I have always called them MLEs, but someone decided to add another letter somewhere along the way. Typically, a single MLE connects to a module, hence the name "module-level."

Microinverters are a fraction of the power output and size of their string-inverter cousins. A single microinverter may weigh no more than a few pounds, with a rated output of about 250 W. The lower weight makes them easier to handle, although, as you can imagine, you would need a lot of them to reach the 10,000 W rating of a string inverter. The microinverter converts DC to AC at the module level and sends it down to ground level.

Power optimizers are sometimes referred to as DC to DC converters and pair with a PV string inverter. Optimizers, like microinverters, are connected to their own module but with far fewer internal components. Instead of converting DC to AC, they maintain a fixed DC voltage across the strings. Since our homes do not use DC, this stable voltage must still be converted into AC by a string inverter. The consistent voltage (300 to 400 V) allows the string inverter to operate at a constant optimal level.

MLEs have the same applications as string inverters—residential, commercial, and utility (theoretically). I don't know of any utility-scale installation using MLE, but commercial installations are well represented. Microinverters and DC optimizers have unique system-sizing methods discussed in more detail later. Before moving on to the nitty-gritty technical stuff, let's step back and—

"Whoa, wait a minute. That's it? Which of these inverter types is the best? What should I be using?"

Oh, I'll get to that later when I talk about product selection. As I said, let's step back and take a look at how this industry got its start, got its start, got its start...

(Begin flashback sequence.)

CHAPTER 4

SOLAR BEGINNINGS

As an industry, solar has been around since the 1970s. But there is a specific time frame when its popularity took off and became mainstream. There weren't a lot of PV inverter or module manufacturers to choose from in those early days. Even into the early 2000s, the market offered only a handful of options, but it did not take long for more selection to hit the market. Many charts show the launch of the "solar coaster," a noticeable spike in the mid-2000s when solar exploded.[1] If you have ever felt the g-force rush of a good roller coaster, you have an idea of what it felt like to be in the industry at that time.

THE DOWNWARD TREND OF SOLAR PRICING

Disclaimer: I hate the sales metric of "dollar per watt." It reduces the science, technology, and engineering, particularly of the solar inverters, into a cheap talking point. But since it is so prevalent, I have to use it. More on this later.

In our industry's primordial beginnings (the 1970s), solar modules were about $105 per watt, and a 10 kW system would have run approximately $1,050,000. Needless to say, systems were relatively small back then. In 1980, the installed cost of solar dropped to $75 a watt. Less than 20 years later (1999), total solar installed in the United States was just under 1 GW. The price of solar modules was about $5 per watt, and a 10 kW solar system cost about $12 per watt installed—modules, inverter, hardware, electrical work, and so on. A 5 kW system would cost about $60,000. Back then, people bought solar out of love and because it was the right thing to do.

Until 2001, the largest contributor to decreasing solar module costs was the increase in cell efficiency. In 2001, prices started falling because of an increase in cell-production capacity. By 2008, the total installed dollar per watt had dropped to $9; by 2013, to $5; and in 2020, to a comfortable $2 per watt.[2] That is a decrease of over 83% since 1999. A 5,000 W solar system costs around $10,000—and that is before rebates and incentives kick in (2021). Even if you paid for this system with your credit card, you could make monthly payments of $350 and pay it off in three years (annual percentage rate [APR] of 15%). A $350 monthly payment is less than most car payments.

Once the low dollar per watt turned solar modules into a commodity, I never understood the value in residential solar-efficiency claims. When a single 160 W module cost approximately $700 back in 2008, it made financial sense to string size that system down to the exact number of modules. But now, since they are so cheap, why not fill your roof up with them?

Bigger projects called for bigger inverters, and, soon, central inverters claimed their spot in this growing landscape. It seemed like everyone attending Solar Power International in 2010 was bragging about a 1 MW project in the pipeline as they puffed on their pipes and drew their suspenders out with their thumbs.

Peter King, SMA America's central inverter field installation expert, invited me to watch the commissioning of such an inverter up in Chico, California. At the time, this model was the most massive central inverter globally, with a rated output of 250,000 W. The bus bars looked like mini skateboards, and the 3 kHz hum of the insulated gate bipolar transistors (IGBTs, the devices that convert the DC to AC) had an ominous, almost deafening hum. Later that day, while sipping flights at the Sierra Nevada Brewing Company, Peter told me the only upside if you accidentally touched those live bus bars would be that you would die before you knew what happened.

THE BEGINNING OF THE SOLAR COASTER

Between 2008 and 2015, there was a massive boom in solar company startups. New "big box" solar installation companies sprang up (such as SolarCity, SunPower, and Vivint Solar) with financing options (power purchase agreements, leasing, etc.) that made it very appealing to own solar. Solar module and inverter prices continued to drop, and banks would back a loan or lease for a five- to seven-year payback on a fully installed system. Soon, the market was flooded with financing models and companies willing to push them. Consumers no longer installed solar because it was the right thing to do but because they could get a break on their utility bill. Solar Power International exhibitions would never be the same. It didn't take long for the cargo-shorts-and-flip-flops-wearing crowd to be overrun by suit-and-tie bankers, financiers, and developers. "Anyone up for a smoke break?" took on a totally different meaning.

THE APEX AND THE FAST RIDE DOWN

Ah, those glory years of solar. For me, they were from 2009 to 2012.

SMA America (my employer at the time) was sitting pretty, my profit-sharing bonus was larger than any reenlistment bonus the Navy ever gave me, and I was consistently training approximately 4,000-plus people a year. The slow downward spiral started almost on cue from a sales colleague who previously predicted this solar coaster would last about another five years, and then the ride would be over. He was damn close.

In 2015, Greentech Media (GTM) wrote an article titled "The Mercifully Short List of Fallen Solar Companies: 2015 Edition." The comprehensive list started with the 2015 casualties, and then went back to 2009 to tally the solar industry's losses over the previous six years. The most notable fall from grace occurred in 2011, when Solyndra filed for bankruptcy. When silicon prices dropped by 89%, Solyndra's modules became insanely overpriced compared to their competitors'. They used different materials in their tubular-shaped modules to create a periodic-table-of-the-elements mouthful—copper indium gallium selenide, or CIGS [$Cu(In_xGa_{1-x})Se_2$]. Despite industry acceptance and increased solar harvest, Solyndra could not compete with the rapidly declining manufacturing costs of crystal-silicon modules. If this company sounds familiar, you probably heard of it attached to President Obama's stimulus program, which gave a $535 million grant to Solyndra through the Department of Energy. Solyndra's Chapter 11 made headlines amidst allegations of wrongdoing and conspiracies. It was disappointing to see such an innovative and promising technology exit the market.

The GTM report is a heavy read. Unless you were an investor or kept up with EU solar across the pond, you probably never heard of most of the companies. But the list does take one down memory lane if you have been in this industry for an appreciable amount of time. Sadly, these lists do not account for the mom-

and-pop casualties who went out of business and left thousands of stranded homeowners with no means of service or repair. I have quite a few acquaintances who make a lucrative living as third-party service providers.

The disappearance of solar companies seemed out of place considering the industry was booming. Installs were going in like crazy! Unfortunately, since there were so many modules and components to choose from, manufacturers slowly lowered their prices to compete. It didn't take long for a race to the bottom. Many manufacturers and installers were providing their products and services *at cost*. That didn't leave much margin to keep the lights on. So they slipped quietly away. Even now, we see casualties. As I prepared this book for final review, a friend of mine texted me an image of his local newspaper with a headline announcing SunPower was closing its SolarWorld module factory. A week after his text, a *Solar Power World* article announced Panasonic was exiting the solar module manufacturing business. Another week later, GTM was no more. Craziness.

THE RISE OF MODULE-LEVEL ELECTRONICS

One of the most influential factors contributing to #SolarForAll was the advent of a commercially available microinverter. Microinverters added a loop to the solar coaster. In 2006, Enphase made history with the first mass-marketed microinverter. However, they were not the first microinverter ever sold. The concept goes back to 1991, when a Dutch engineer, Henk Oldenkamp of OKE-Services,[3] designed and produced the grandfather of modern microinverter technology. Henk's microinverter reached 100,000 installations by 1999 (mostly in the Netherlands), but the concept just never made it to the United States.

In 2010, I told one of the German SMA executives that in my

training classes, more people were asking about these small invert-ers installed underneath the modules. They responded,

"Ah, yes. Enphase. They have only a 3% market share. Microin-verters make no sense."

If you reread that using a thick German accent, it adds more impact, considering that statement did not age well. Despite a few reconciled product-quality hiccups along the way, which are common among most companies introducing new products, Enphase rose to 19% market share in 2012 and jumped to 37% in 2014 (GTM estimates[4]). Enphase even took over an entire state. Over 80% of all solar installations in Hawaii belong to them.

Many installers migrated toward the micros; these new module-level devices disrupted the decades-old way solar was sold, designed, and installed. Some tried them and went back to string inverters, some used both, and some left the string invert-ers behind forever. Desperate to hold market share after ignoring the upward trend of microinverter popularity for too long, the leviathan string-inverter companies who once ruled with a loyal following were struggling to stay in the black.

Microinverter and optimizer manufacturers sprouted up almost overnight in hopes of capturing their piece of the MLE-market pie. Even some string-inverter companies released a microinverter to regain some of their lost residential market share.

SMA America released its first microinverter in 2013 after two-plus years of development. Spirits were high; after all this time, we would have a competitive product! When it arrived at our Rocklin office, one of the product managers called me in to have a peek at it. We were both appalled.

The first rule of product development is you must first be equal to the competition, and then you must show how you are better. Otherwise, what would be the incentive for customers to swap manufacturer mid-stroke? The market rejected SMA's concept, and the fancy robot that assembled the microinverters sat silent for too long. Even if the idea did eventually take off, it would cost an obscene amount of money to recalibrate the automation line. Lesson learned.

Most string-inverter companies stuck to what they were good at and focused on lucrative commercial and utility-scale PV, where microinverters and optimizers may not be an appropriate solution. I will justify this statement later.

The microinverter revolution was a fantastic transformation. People who were raising hogs three months prior could now get into solar and install microinverters because it was that easy. Unfortunately, a lot of the systems looked like former hog farmers had installed them.

The string-inverter companies that saved the industry from the brink of collapse back in the early 2000s were now mocked for their lack of module-level vision. Some MLE installers, who had once been scared of solar, got cocky and bold now that they had a "solar for dummies" product.[5] I remember a North American Board of Certified Energy Practitioners (NABCEP) event (circa 2011) where an installer who had never even installed a system challenged me on the SMA string inverters. Standing in front of a full classroom, I calmly refuted his claims, parroting the sales and marketing literature. But it didn't stay that way for long.

"Why does SMA only offer a 10-year warranty instead of a 25-year warranty? If you are so good, give us a better warranty."

I asked him if he thought it odd that MLE companies that had only been around a few years offered a 25-year warranty, while established string-inverter companies with 20 years of expertise provided a 10-year warranty.

The climax came when he said SMA would not stay in business with our "dinosaur" string-inverter products. I retorted something about how his business model should include removing and replacing all those MLE components that will either fail or wear out over time. He frustratingly snapped back with, "I don't even know if I will be in business in five years!" I advised him *not* to install SMA and pointed him to a microinverter class just down the hall. I almost made a snide comment about his future as a hog farmer, but I refrained. He and his buddy got up and left. Great vibe for the rest of that class!

Ben Castillo, a fellow SMA trainer and friend, once listed four things you never bring up during training events: politics, religion, favorite breakfast cereal, and micro- versus string inverters. It got ugly for a while on social media as string-inverter apologists challenged the microinverter claims. My LinkedIn views exploded. Just as quickly as the online battle had started, it was over. People moved on and stopped trying to convince those who could not be convinced, at least in those forums.

NEW TECHNOLOGY EMERGES

In 2011, a different type of MLE architecture aligned with the microinverters in the marketing fight against string inverters— DC optimizers. Optimizers boasted the best of both worlds: module-level control but with far fewer parts than microinverters, all feeding a more efficient string inverter. Although there are other power-optimizer companies, SolarEdge and Tigo dominate the market for this architecture.

The MLE sales and marketing approach hit the US market with brute force, and their appeal quickly grew:

- Small mom-and-pop solar installation companies could now break into this new market with a simple product that almost sold itself.
- The 20-year warranty was their sales pitch ace up their sleeve (and still is).
- Designers didn't have to use complicated string-sizing programs.
- Large, wholesale electrical distributors had fewer part numbers to worry about and enjoyed less demand on their shelf space.
- They offered shade mitigation for partially shaded rooftops.
- There was no high-voltage DC coming down from the roof, which meant using lower-skilled workers for installation.

These are legitimate reasons to use optimizers and microinverters; however, the legitimate intertwined with marketing *Kuhscheisse*. In the "old days," installers refrained from putting solar arrays on roofs that would be partially shaded between 9:00 a.m. and 3:00 p.m., typically referred to as the solar window. But this strict rule of thumb was a salesperson's nightmare since it would deny them sales for a significant number of residential applications. Many salespeople still sold systems that had no business being on a roof, but a micro or optimized installation would at least help them sleep better at night.

I will never forget when, at a Waikiki training event, I looked off the patio down to a convenience store 20 floors below and saw that all four roof sides of the store had solar on them. One side had a palm tree lying *on* the array. It was apparent the installers moved the palm fronds out of the way to install that section of modules. Here it is in all its glory.

I understand and support the use of MLE (or string inverters with shade-tolerant maximum power point tracking [MPPT]) on these partially shaded areas from a business perspective. Nobody wants to turn down a job, and something is better than nothing. As long as everyone is happy with the production estimates, then full speed ahead. But regardless of PV inverter topology, systems like this one (I've seen much worse) do not flatter our industry.

WRAPPING IT UP

The number of solar equipment manufacturers has steadied out in the later part of the 2010 decade, and they each jockey for position on the quarterly market-share charts. As of year-end 2020, Enphase and SolarEdge own 80% of all solar installations in the United States—a sobering testament to their market acceptance. Promising technologies continue to challenge the status quo. And that brings us to the bottom of the roller-coaster track, but there are more twists and turns ahead, like any good ride. Now begins the storage coaster! OK, that's not as sexy as "solar coaster," I admit. Still, the same patterns are emerging:

- New storage companies and battery manufacturers are popping up with bold product claims.
- Installers and consumers are wading through sales and marketing materials, wondering what to do.
- The NEC is (slowly) changing to adapt to the new landscape.
- Industry rebates and incentives are on the ebb and flow.

If you want to learn more about the solar and storage industry's beginnings, I highly recommend the short film *Solar Roots: The Pioneers of PV.*[6] It contextualizes the importance of these early days and immortalizes remarkable and knowledgeable people to whom the industry owes its success and gratitude. Jeff Spies put a lot of time and effort into this film, and it is worth watching.

Considering how fast things change in this industry, it will be interesting to see how well this book ages in 5 to 10 years. Away we go!

CHAPTER 5

INTRODUCTION TO STORAGE

Batteries have been around almost as long as the solar cell. Battery history begins the same way all inventions do—a cascading impact of discoveries that got us where we are today. Founding father, inventor of the printing press, and discoverer of electricity Benjamin Franklin is credited with creating the first (sort of) battery in 1749. Franklin even invented the word "battery." I wonder if the townsmen, jealous of Franklin's many successes, thought, "Jeez, man, save some three-penny uprights for the rest of us!"

At the turn of the nineteenth century, Alessandro Volta invented the first true electrochemical cell. Volta used copper and zinc with an electrolyte composed of sulfuric acid and salt water to demonstrate current flow. In 1859, French physicist Gaston Planté invented the first lead-acid, rechargeable battery. The continued development of battery technology paved the way for battery deployment in equipment and vehicles in the early 1900s. Skip ahead to 1993, when Sony released the first commercially available rechargeable lithium-ion battery. We've come a long way.

Mainstream residential solar started to appear on the annual GTM charts around 2008.[1] A few bats of the eye, and in Q4 of 2013, residential storage exploded.[2] The industry has been breaking quarterly storage-deployment records since 2018. It is so popular "storage" is considered a barbaric description. We needed more syllables to describe the same thing, so "energy storage systems" (ESSs) became the hip, new phrase.

The COVID-19 pandemic caused a noticeable dip in storage's Q2 2020 charts, but growth projections remain stable for this multibillion-dollar industry. Lithium storage is all around us now—in cars, our homes, and utility infrastructures. There will be a time in the not-so-near future when kids will look at internal-combustion-engine vehicles with the same curiosity as people looked at the electric vehicle forerunners.

The mere mention of stationary battery storage, even up to the mid-2000s, used to conjure images of hunting cabins powered by piecemealed solar arrays, hippie-engineered solar inverters,[3] golf-cart batteries, and crude charge controllers made out of RadioShack parts (Google it, kids). It's not like that nowadays. Storage power electronics matured as fast as computer-chip technology. Magnum Energy, OutBack Power, SMA Solar, Trace Solar, Xantrex, and others forged the way through the utility-controlled gauntlet to bring us where we are today. Smart, highly efficient, and durable products flood the market with a hungry-locust appetite for all storage applications.

In the mid-2010s, industry professionals foretold a day when installing solar without storage would make no sense. Germany was the first to prove this when they gradually reduced the solar feed-in tariffs, and the United States soon followed. Some markets beat our projections. Hawaii and Puerto Rico, island states with a massive amount of PV penetration into aging grid infrastructures, adopted strict interconnection requirements that forbade the backfeed of excess solar into the grid. This new requirement caused a lot of commotion. PV inverters generate power as soon as there is enough sun, like a waterwheel spinning when the creek starts flowing. The new technology would throttle down solar production to prevent any excess backfeed—and it made homeowners mad! They paid a lot of money for a system that, by design, hobbled itself. All that sunshine was going to waste.

It didn't take long for storage to become the saving grace for these areas. Instead of throttling down the PV, batteries stored the excess electrons for later use. Huzzah! High PV penetration wasn't the only reason storage started to gain popularity. Some states facilitated the adoption of storage through increased time of use (ToU) rates or peak charges. As of 2020, Arizona has the highest number of people "voluntold" to participate in a ToU rate structure. In 2016, for the first time in history, storage had a legitimate return on investment (ROI). Not as respectable as the coveted 5 to 7 years for a solar system, but anything under 10 years is still pretty good.

During critical times of day, like when people come home and start turning on their air conditioning, grid support usually requires peaker plants to quickly ramp up electricity generation. The peaker plants are quite dirty to operate and release significant amounts of pollutants. Over 1,000 fossil-fuel-driven peakers in the US average yearly run times of about 300 hours.[4] These peakers are also very costly to operate. For example, New York City has 46 such plants that cost 1,300% more to produce peaker power than the average electricity cost. We can do better than this.

Storage applications have evolved to give new meaning to the term "smart grid." Utilities use centralized mass storage facilities for grid services, including operating as a replacement for fossil-fuel peaker plants. The United States boasts the most extensive utility-scale storage facilities in the world. Ranked by capacity (2019):

1. 2.5 TWh Ravenswood project in New York City (yes, terawatt-hours)
2. 1,200 MWh (megawatt-hour) and 400 MWh projects by Vistra Energy at the Moss Landing facility in Moss Landing, California
3. 900 MWh Florida Power and Light facility.
4. 730 MWh system by Tesla at the Moss Landing facility (Moss Landing is a pretty popular place for energy storage!)

There are also many lithium storage sites in the 20 to 40 MWh range. This type of storage is also catching on worldwide.

Instead of large-capacity, centralized storage systems, imagine having hundreds or thousands of distributed storage systems across hundreds of communities ready to provide grid support at a moment's notice. This type of grid support is not some mad scientist's renewable-energy reverie but rather a reality. Germany's intelligent-energy storage manufacturer, sonnen GmbH, has the world's largest active residential virtual power plant (VPP). This countrywide VPP is called the sonnenCommunity and has been running since 2015. Not only can the homeowners' energy be used for grid services, they can also supply each other with surplus energy—no small task, even for a country the size of Texas. In the United States, sonnen, Inc. partnered with a local utility company and a nationwide real estate developer to create the nation's first flagship project for a behind-the-meter solar-plus-storage project. *Utility Dive*'s Project of the Year (2019) was awarded to the Soleil Lofts complex outside Salt Lake City, Utah.[5] This is a true smart grid, with dispatchable distributed energy resources (DER), and is actively using storage to provide grid services. It is the future!

Many people talk about VPPs today in the same way people talked about their 1 MW solar projects at the 2010 Solar Power International conference in Los Angles. While there may be theory crafting, hand wringing, and speculation concerning VPPs, not many companies are doing it. But some are aggressively pursuing this cash cow. Australia-based company Solar Service Group is on track to "host the world's largest virtual power plant" by mid-2021.[6]

"You gonna talk about how these things work or what?"

Absolutely! Let's begin with some common battery terminology and a basic understanding of how a battery works.

"No, no! I meant like the different types I keep hearing about… not more of this sciencey cra—"

Too late!

BATTERY TERMINOLOGY

Most of us have been exposed to battery jargon, although we might not know what the terms are or even mean. For example, most people have a cell phone, and most of us get a knot in our gut when the battery percentage turns red, right around the 15% to 20% level. We understand we must now plug our phone into a socket using the inconveniently shaped outlet adaptor. We leave the phone plugged in until we return and see the percentage at 100% or a comfortable level that allows us to run out of the house to our next appointment. You have experienced these battery behaviors countless times but probably never thought about connecting the dots. Let's briefly cover some storage terminology so we can all be on the same page.

Lead-acid—A widespread battery type that uses lead plates and electrolytes to store energy.

Lithium-ion—A battery technology using lithium ions that move between an anode and a cathode using an electrolyte as a conductor. Sometimes shortened to LIB or Li-ion.

Capacity—The amount of energy the battery bank can store. Analogous to the gas tank in your car. Capacity is measured in either amp-hours (Ah) for lead-acid batteries or kilowatt-hours for lithium-ion batteries.

Charging—The act of "filling up" a battery using an electrical source.

Discharging—The behavior of the battery when it is providing power to your device or home.

State of charge (SoC)—A percentage that shows how much battery capacity is left. This is what your cell phone is telling you.

Depth of discharge (DoD)—A percentage that tells us how much battery capacity we have used. It is the inverse of SoC. There are likely only two probable places to see the DoD: when reviewing a lead-acid battery spec sheet or looking at a lithium-ion battery sales and marketing sheet. Lead-acid has used DoD for as long as anyone can remember. Companies usually only refer to the DoD when describing how much of the battery capacity you can use as a marketing and warranty feature.

A phone battery displaying 25% means there is one-quarter of your battery life remaining. If we were to express the percentage using DoD, it would be 75%, meaning we have used three-quarters of the available battery capacity. Are you a glass-half-empty or a glass-half-full person?

State of health (SoH)—A percentage that expresses how much battery life is remaining. An SoH value of 70% to 80% is end of life for most battery types.

Cycle—The act of discharging and charging a battery. Cycling a battery has a different effect on battery life depending on the battery type. Typically, lead-acid batteries do not like regular cycling past 50%. Their cycle charts show cycle versus DoD, and you will get more cycles the less frequently and the shallower you take the discharge. In contrast, lithium-ion batteries can cycle much lower, 90% to 100% of their usable capacity. Most battery manufacturer 10-year warranties guarantee between 2,000 and 10,000 cycles. You will rarely see a cycle versus DoD chart for lithium-ion batteries—a hard thing to overcome if you have been around lead-acid batteries.

Lead-acid counts a cycle much differently than lithium. Any discharge and charge, regardless of depth, is considered a cycle for lead-acid batteries. Conversely, lithium-ion battery-management systems do not count a cycle until it measures 100% discharge and charge. For example, let's assume you routinely used only 20% of your battery capacity each day. A lead-acid battery system would count each 20% day as a full cycle, while a lithium-ion system would not count a full cycle until five days had passed.

That wraps up this chapter. Now let's dive into each of the battery types in more detail. Onward!

CHAPTER 6

BATTERY TECHNOLOGY OVERVIEW

Battery technologies are as essential as the types of PV inverter technologies. The same principle applies—there is no one size that fits all, and each has its strengths and weaknesses. Although tempting to take these pros and cons and visualize them as a chart, instead, it will be much more fun to talk about them. Well, I guess "fun" is subjective. My cousin Karen assists neurosurgeons in performing brain-stent surgery. Stents are one-twelfth replicas of lightning rods inserted into the brain that direct small electrical currents between them. They have proven effective in treating aneurysms and other brain disorders. Karen gets as giddy as someone watching a *Housewives of <insert city>* reunion episode when talking about minimally invasive endovascular stent-electrode arrays for high-fidelity, chronic recordings of cortical neural activity. I just nod my head and keep drinking my Hefeweizen.

There are scads of promising battery technologies, but I will focus on the two most popular types used in residential storage: lead-acid and lithium-ion. I will spend only a little time talking about lead-acid since the trend is shifting toward lithium-based battery technologies. Perhaps new technology will overthrow lithium-ion as abruptly as lead-acid fell to lithium, but it won't be for a while.

Lead-acid has been the de facto battery technology for off-grid and residential storage for decades. It is battle-tested, affordable, and highly recyclable. Almost all lead-acid material is recyclable—around 98%. It is still an attractive battery type for developing countries with hundreds of thousands or even millions of people without power and too far away from an electrical source. Solar plus storage is their only hope.[1] Many people still prefer this technology to lithium.

Lead-acid battery design hasn't changed much since Volta's time. However, there have been leaps in material improvements that boost energy density, cycle count, and reliability. Lead-acid construction consists of lead plates submerged in an electrolyte. These tried-and-true batteries are all around us. Manufacturers tweak the chemistry and plate construction to fit the desired application. Batteries for cars and boats have thin plates to provide quick-cranking amps compared to the thicker plates in deep-cycle batteries. As the name implies, deep-cycle batteries can happily withstand deeper discharges, but this superpower comes at a cost. Frequently abused by electric vehicles, PV systems, uninterruptible power supplies, remotely located cell phone towers, and even submarines, deep-cycle batteries typically have lower cycle counts and degrade much faster than other lead-acid batteries. You never get something for nothing.

As I said before, residential grid-tied energy storage is shifting to a lithium-based solution.[2] To all my die-hard lead-acid friends and users, I sincerely apologize. This is when the old-timers, of whom

I know many, will rant and rave about their beloved lead-acid batteries. And for good reasons! Lead-acid is still the battery of choice for off-grid applications. It is a much cheaper alternative than lithium and can take a beating. But it is generally not a good fit for homeowners who do not have the space for lead-acid and do not want to deal with the babysitting these batteries require.

Many installers have a dismissive attitude toward lead-acid because of the design complexity and periodic maintenance necessary to keep the batteries healthy. They do not want to enter into that kind of service commitment. Historically, most solar companies would not entertain a storage project unless they had to. That's right. They turn down money because it is too much of a pain for them. But that dismissiveness has come to a screeching halt thanks to lithium-ion.

CHAPTER 7

LITHIUM-ION BATTERIES

Lithium-ion batteries have been around since the 1970s, when the oil crisis prompted a chemist at ExxonMobil to begin researching the idea for a better rechargeable battery. The dream was that, one day, stored energy would allow us to tell our Middle Eastern oil overlords to kiss our lithium-based asses and reduce our reliance on fossil fuels. Who would have thought it would be a combination of lithium and silicon that would pave the way for that future?

As with all initial research, the first lithium battery designs were a bust. The combination of titanium and lithium in Exxon's design resulted in fabulously explosive failures and ultimately led them to abandon their research. It took a while, but in the late 1980s, researchers developed a safer prototype with substantially better stability, efficiency, and energy potential than previous iterations. The key was to replace the metal lithium anode with a carbon-based element and the titanium cathode with cobalt. Voila! Now almost every handheld device would be safely powered by a lithium-ion battery. In the mid-1990s, Sony paved the way for the infancy of the electric vehicle revolution.

"Wait. Hold up. You just said they got rid of lithium. So why do we call them lithium-ion batteries?"

Manufacturers inject lithium ions into the battery mixture. However, there are batteries with metallic lithium and even titanium, but they use different materials to stabilize the cell. It is all about finding the right mixtures.

Lithium-ion has opened doors for the layman in the same way microinverters opened doors for the solar skittish. These battery systems require little periodic maintenance, are easy to install, are more cost friendly, and have a certain sex appeal. Lithium-ion has taken over the residential market. Its acceptance for new-home construction as a standard feature reveals a promising glimpse into its future. In May of 2019, battery experts gathered for a three-day conference in Berlin. Their findings concluded "that the current vibe about the need of future technologies after the lithium era and, thus, the quest for which new technologies can replace lithium-based battery technology, are somewhat inappropriate and misleading (partially incorrect), respectively."[1]

If you haven't worked for German companies as long as I have, allow me to loosely translate that last paragraph: "STFU about all these other battery types until they are a proven technology and economically viable for mass production." The paper lists an impressive number of promising technologies, including flow, sodium-ion, metal, and metal air-based batteries. According to these experts, none of these technologies will replace lithium-ion anytime soon and battery scientists shouldn't be spending too much time trying to replace lithium-ion. Instead, they should develop these promising technologies for either hybrid applications with existing lithium technology or as stand-alone solutions for their appropriate applications. I also recognize and bring attention to the fact that this scientific, peer-reviewed paper used the word "vibe."

LITHIUM-ION CHEMISTRIES

This section may not make you an expert in lithium technology, but you will glean enough jargon to become the life of the party. Lithium-based articles and scientific papers love to string periodic elements together to describe various substitutions that make up the battery. Here is an example from a paper about recycling lithium-ion batteries:

"The most common being $LiCoO_2$ (LCO), $LiFePO_4$ (LFP), $LiMn_2O_4$ (LMO), $LiNi_{1/3}Mn_{1/3}Co_{1/3}O_2$ (NMC-111, abbreviated as NMC), and $LiNi_{0.8}Co_{0.15}Al_{0.05}O_2$ (NCA)."[2]

How far did you get before you thought, "Why am I reading this?" Not to diminish the importance, because chemistry matters, but to the layman and the NEC, it is (mostly) irrelevant. But my last statement is heavily debated with frothing enthusiasm.

There are two primary chemistries in the residential stationary storage market. Chemistry refers to the stuff inside the battery, and "stationary" describes the application opposite of electric vehicles. However, there is an exciting crossover between stationary and electric vehicle batteries. Battery chemists are always trying new ingredients to boost lithium-ion efficiency and energy or power density. Although there are five or six kinds of lithium-ion chemistries (depending on the source), only two have made it to residential storage big-time. These two noteworthy chemistries are LFP and NMC. The chemistries use abbreviations since using the periodic table's names for them would be too exhausting. Since we can't be bothered to type out "be right back," we are certainly not spelling out the chemical designators for lithium iron phosphate and nickel manganese cobalt, respectively.

I describe lithium-ion as the make of a car, and the chemistries are the models. So let's kick the tires!

Battery life expectancy is an important consideration since these systems represent a considerable investment. People want to know how long they are going to last. Unfortunately, there isn't much empirical, field-tested data for residential lithium-ion ESSs since these things haven't been around that long. Sales and marketing literature boasts high cycle counts and long warranty periods, but warranty does not equal reliability, as I've said many times. So we have to lean on laboratory-accelerated life-expectancy testing and some serious math with a dose of science to help put our minds at ease.

A 2020 study, "Lifetime Expectancy of Li-Ion Batteries Used for Residential Solar Storage,"[3] compares and contrasts LFP and NMC chemistries. The research focuses on the results of what the authors describe as state of the art (SoA) batteries, and two well-known manufacturers each represent a chemistry. Although the study uses three real installation sites in different parts of the country, the country is Italy. Europeans are far more energy conscious than Americans, so they tend to need smaller systems. The numbers in this study will not represent the vast majority of US residential grid-tied systems. But that's OK! Remember, trons are trons.

TAKEAWAYS

- LFP capacity degrades faster than NMC in the first year of operation, but then capacity fade evens out.

- Larger-capacity systems presumably last longer than smaller ones since they are not cycled as often or as deeply.
- Overall, LFP batteries appear to last longer than NMC: 20.88 versus 17.75 years for a 10 kWh capacity system.
- Larger-capacity systems (greater than 10 kWh) benefit from larger PV systems.
- Larger systems had a 15% greater life expectancy than smaller systems.
- Battery-location temperature affects battery longevity more than any other variable.
- Load profile did not have a significant effect on battery degradation. (I would have bet money that it did.)

Money shot: high-quality, state-of-the-art batteries have a life expectancy well beyond the 8- to 12-year estimated break-even point (ROI). The authors concluded these systems would reach their warranty periods. This sentiment also seems to jive with the annual DNV Battery Scorecard.

I think this study puts to rest any concerns about warranty periods, right?

"OK, cool story, bro. But why would one ESS manufacturer use NMC and another use LFP? Why don't they all use the same thing?"

The Tesla Powerwall and the LG Chem RESU are the only two mainstream products (as of this writing) that use NMC batteries in the US market. These two products use electric vehicle batteries. It was a no-brainer for them: the economy of scale was already in place.

Radar charts for NMC and LFP show the distinct advantages of each.[4] Pricing between the two battery types appears to be equal, but LFP is typically lower. While NMC has better specific energy, LFP has higher safety and longer life-span dimensions. I am not advocating that NMC batteries are dangerous, only that LFP is safer. Let's compare two cars: Car A has seat belts. Car B has seat belts, driver-side airbags, and traction control. While both vehicles are safe, Car B is safer.

To show battery response to a puncture, I used to show a video in my classes. The NMC puncture produced an arc and subsequent fire. The LFP battery heated up but did not explode. A quick internet search for "NMC LFP puncture" will show you what I mean. But I stopped showing that video since the NEC, local inspectors, building codes, homeowners, and so on, make no distinction between battery chemistries. To them, USDA Prime, Choice, and Select are all the same. Gag.

Additionally, these videos do not account for electrical, mechanical, and software safeguards within the systems as a whole. Unfortunately, ESS and electric-vehicle fires have scared people to the point where safety codes require outdoor installations. However, garages are not considered "indoors," for now. Good thing, since most homeowners would throw a fit if they could not use their garage to park their electric vehicle.

NMC CHEMISTRY

NMC batteries use these exotic metals to help increase either power or energy density. Manufacturers will tweak the chemistry depending on the application. Typically, an NMC battery is one-third nickel, one-third manganese, and one-third cobalt. Battery chemists refer to this mixture as a 1-1-1. It works well in applications or devices that frequently cycle the battery: electric vehicles,

ESS, power tools, gadgets, e-scooters, and the high-powered laser pointer I use to tease the cat. Sometimes, manufacturers will use a 5-1-1 or 3-1-1 to increase either energy or power density, depending on the application and environmental conditions. But they will not modify the mixture for different electric vehicles. This is the NMC Catch-22: drivers want their electric vehicle to perform just as well in the New Mexico desert as it does in the Northeast's bitter winters. Unfortunately, electric vehicle batteries are one-size-fits-all for now.

GEOPOLITICS OF NMC BATTERIES

Along with lithium, manufacturers source the other metals in NMC batteries from all over the world. Some of these areas are more stable than others. A conflict in just one of these areas could drastically affect the supply chain for NMC battery production.

COBALT

The Democratic Republic of Congo (DRC) provides the world with 60% of its cobalt, which could jump to 80% by 2025.[5] Although NMC battery manufacturers are developing ways to reduce the amount of cobalt needed, forecasts show an increasing demand for this element. Hopefully, the good people of the DRC will benefit financially from this demand. About 50% of all mined cobalt goes into lithium-ion batteries. The switch to LFP means less cobalt as a manufacturing material, which means costs could come down. This seems to be a growing trend.[6]

LITHIUM

Most of the world's lithium comes from Argentina, Australia, and Chile. China has a lot of control over converting raw lithium into a usable product, as it houses two of the world's largest facilities. Lithium is a $3 billion a year market. However, prices should drop since its supply exceeds demand—good news for electric vehicle and ESS manufacturers!

MANGANESE

Ninety percent of all mined manganese goes into steel production. The remaining 10% is divvied up between water-treatment filters, fertilizers, multivitamin supplements, yellow cake (both uranium and the dessert), and rechargeable lithium-ion batteries. Holding the title of the fifth-most abundant element on our planet for the last 4.6 billion years, most mined manganese comes from South Africa.

NICKEL

This element is slowly replacing cobalt in lithium-ion batteries. The explosion of NMC battery manufacturing in recent years has caused the price of nickel to surge. About 40% of global supply comes from Indonesia and the Philippines, and the market is worth over $30 billion annually. It might be time to shift some of your commodity stocks from pork bellies to nickel.

LITHIUM IRON PHOSPHATE CHEMISTRY

This battery type is the most predominantly used in residential storage applications. Electric vehicle manufacturers experimented with this chemistry, but it was short lived, and they quickly switched to NMC. The iron proved to be a detriment. LFP does not have the exotic metals of NMC, and iron and phosphates are readily available. Interestingly enough, some electric vehicle manufacturers are considering going back to LFP for safety reasons.

RECYCLING (DEBBIE DOWNER ALERT!)

There are not many places that recycle batteries, and those that do are expensive and may not even recover some or all of the precious metals—this is true for stationary and electric vehicle batteries. Roughly only 5% of lithium batteries are recycled,[7] and a lot of the recovered lithium goes to lubricants. Cobalt is the most precious of the metals, but recovery is hopelessly challenging financially. I foresee government subsidies in the future to help with this.

There are several recovery processes. Most require immense amounts of heat, which needs a lot of energy to produce. I'm sure someone has calculated the recovery benefits versus mining and production costs weighed against the environmental impact, but I am reluctant to read it. Regardless, recycling as much as possible probably beats batteries rotting in landfills.

Another barrier to recycling is the manual process it takes to disassemble an electric vehicle battery, in addition to the safety hazards of being around 300 to 400 V. Disassembly automation is another challenge. Every electric vehicle manufacturer uses a different battery pack, which means the recycler must use different robots to handle the various sizes, bolt locations, and battery placement configurations. There is hope for some batteries. Apple has taken charge of its battery recycling by making it automated. Their robots can disassemble a cell phone in seconds so the recyclers can recover the precious metals.

Stationary-storage recovery suffers the same limitations as electric vehicle batteries since the battery packs come in all shapes and sizes. Deep down, in places we don't want to talk about at parties, we have to come to grips with the fact that very few of these residential batteries are recycled. The upside is that these systems are relatively new and haven't hit their end-of-life cycle, which means there aren't many to recycle anyway. Perhaps by the time these systems reach end of life, we will have an efficient and effective way to achieve mass recovery. There are companies worldwide trying to beat these challenges, with hope-inspiring results. A joint UK and US group determined "comprehensive labeling, simplified solid-bulk architectures, easy-to-open design, and reversible adhesives and binders" would solve the overwhelming majority of recycling barriers.[8]

Now that all that is out of the way, let's get into how the batteries work with the rest of the system.

CHAPTER 8

HOW IT FITS TOGETHER

We've covered individual components' basic operations in previous chapters, but fitting the pieces together is not always easy. There are many ESS manufacturers, and it is not feasible to cover each one individually. Different architectures require different connecting means and may have a slew of balance-of-systems components. However, they all share the same basic features and operational characteristics.

As I said, the internet has shrunk the gap between novice and expert. Anyone can use their favorite search engine and find "Brand X installation manual" or "How does Brand Y work." Manufacturers offer vendor training, and I highly recommend sitting in on some. Correction, I highly recommend leaving your phone off, your laptop screen down, and sitting through the training. I see so many salespeople get up to take a call or sit there answering emails during training sessions. Why give up a few hours, or even a day, of your time if you aren't going to learn anything?

ENERGY STORAGE AS A COMPLETE SYSTEM

Most ESSs are connected in similar ways but may use various components not shown below. Energy flow is bidirectional depending on solar production and the battery operating mode. The following diagram is an oversimplification of the components, but we are not submitting this for authority having jurisdiction (AHJ) approval. It conveys effect, not realism. This diagram features a sub-panel referred to as a "protected load panel" (PLP). The PLP contains the breakers directly fed by the solar and storage system if the grid goes down. This sub-panel has many names: backup panel, essential-loads panel, and critical load panel.

CRITICAL LOAD PANEL RANT

Although calling a module a panel is just a minor annoyance, referring to the sub-panel as a "critical load panel" could be costly. First, it is a no-no by the strictest definition of the NEC. Several articles of the NEC discuss critical systems (700 and 708 for starters). These critical systems include emergency egress lighting, fire detection and alarm systems, elevators, and even hospital life-support systems or heating, ventilation, and air conditioning (HVAC) in commercial buildings.

Second, although engineers and homeowners have different opinions on critical devices, most professionals agree most homes do not contain these critical loads. But some do, and this is where things can get messy.

I recall an installer who told a homeowner that a "critical load panel" would be installed to keep those loads running when the grid went down. On commissioning day, the homeowner and the installer were in for a big surprise when they cut utility power to the house to test that backup capability. The homeowner, who was in the medical industry, was horrified when his mother's oxygen system shut off and required a restart.

I was on a speakerphone call with this installer and the homeowner. I explained that most residential ESSs would not switch over quickly enough for these critical devices. Do you know how hard it is to try and juggle improper terminology with maintaining installer credibility? Not fun. The installer ended up paying for a small uninterruptable power supply (UPS) to cover the switchover time.

Back to our power flow diagram.

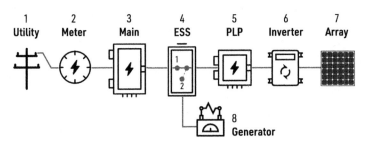

Utility power (1) enters the home through the utility meter (2). The utility power feeds the circuits and loads on the main panel (3). It then goes through a large breaker into the ESS (4). All ESSs have a grid-isolation relay or a transfer switch. In normal grid-connected operation, the relay will be in Position 1. When grid power is lost, the relay will switch to Position 2 and isolate the microgrid. The power then flows from the ESS into the PLP (5), where it connects to another breaker. The solar inverter (6) connects to the PLP with its breaker. Microinverters will, of course, be underneath the array (7) but will also connect to the PLP using a breaker. Although not always used, but highly recommended, the backup generator (8) connects to the ESS. Once the sun comes up, the magic begins! The power flow reverses direction depending on the amount of solar production. Let's start on the right side of the diagram.

Once enough sun is shining for the PV inverter to start producing, we have a reversal of energy from right to left! Generated solar (6) flows into the PLP (5) to feed devices, appliances, and other essential loads—like our gaming consoles and beer fridges. Any excess PV will backfeed into the ESS (4) to charge the batteries.

Achtung! The ESS only uses excess PV production to charge the batteries. If the solar is producing 1,000 W but the microwave is reheating leftover pulled pork, the microwave will consume all excess solar. None will go to the batteries.

Once the batteries are full, any excess PV will backfeed to the main panel (3) to power the house loads. If there is still excess PV, it will go into the grid (1) through the all-knowing utility meter (2).

This architecture is an AC-coupled design, meaning the ESS uses directly converted PV inverter AC energy to charge the batteries. DC coupling designs use raw PV to charge the batteries. However, it must still be converted to AC to power our devices and appliances. I will compare and contrast these two architectures throughout the rest of the book.

OPERATING MODES

ESSs use three different operating modes that affect their charging and discharging behaviors. There are many considerations that impact which operating mode is best for the customer.

BACKUP

In the most operationally boring of the three modes, ESSs will just sit there with a full charge, patiently waiting for the grid to go down. Sure, the battery-management system will trickle charge the batteries to keep them healthy, but the real magic doesn't happen until there is a loss of grid power. The system will switch over to backup mode, provide power to the house, and use solar to recharge the batteries. An internal or external automatic transfer switch or a transfer relay inside the ESS will now be in Position 2 so the microgrid is isolated from the utility. Refer to the previous graphic for power flow from right to left.

Winters can be brutal, with consecutive months of cloudy weather. There will likely be times when the solar production will not support powering loads and charge the batteries. Northeast winters will almost always necessitate a backup generator to charge the batteries when the solar cannot. Systems in the mid-South might need a backup generator when a freak arctic cold front slips through and cripples the grid.

"Wait, what happens if the grid is out, the batteries are full, and there is still excess generated solar? Since the transfer relay will not allow backfeeding into the grid, where does that excess power go?"

Engagingly perceptive, Mr. Holmes! Since ESSs cannot backfeed solar to the grid during a grid outage, the engineers had to get creative. When those three conditions exist, some systems will activate smart plugs or other home-automation devices to start using this excess solar. In the off-grid world, devices that consume excess solar are called dump loads, a fancy name for a heating element. Some systems can throttle down the solar inverters using their proprietary methods. This software feature is an elegant and fast way to stop PV production. DC-coupled systems typically have no problem throttling down PV production since they are on the battery's DC side. Others must use frequency shifts to turn off the PV inverter(s).

All PV inverters in North America must operate between 59.3 Hz and 60.5 Hz, as required by UL 1741. Once the grid goes down and the battery systems create a 240 V / 60 Hz microgrid, the PV inverters recognize this in-spec "grid" and resume making power once their five-minute mandatory wait time has expired. Once the SoC is near 100%, the battery inverter will increase its microgrid frequency above the 60.5 Hz threshold. The PV inverters will disconnect from the high out-of-spec grid and start the five-minute reconnection attempts. The loads are now running solely on battery power, and the SoC will begin to drop.

Once the SoC drops to the manufacturer's preset limit, the battery inverter will decrease the frequency back to 60 Hz. The solar inverter will see the frequency is back in spec and start making power after waiting five minutes. Rinse, repeat.

SELF-CONSUMPTION (OR SELF-SUPPLY)

This mode favors systems that can run on self-generated solar during the day and use the battery system at night to feed house loads. Homes that can operate like this are in a "virtual off-grid" environment. They are self-sufficient and will only sip off the grid when they do not have enough solar or storage to cover loads. They use the grid as a backup generator in case they use too much of their battery capacity. I have seen systems that were so perfectly sized the homeowner could flip the main breaker to the house and run solely on self-generated power. Sadly, these systems are the outliers in the United States.

Some homes can sustain this self-consumption mode through-out the year, but most will only maintain it during the summer months, when the sun is plentiful. Otherwise, they will sup-plement their consumption using the grid. However, this mode never uses the grid to charge the batteries, only solar. The ratio of self-consumption versus grid usage is usually displayed as a percentage. The higher the percentage, the less reliant you are on the utility. I like to call this "grid independence."

The Investment Tax Credit (ITC) is one of the most influential subsidies for solar. A homeowner can get up to 26% of the cost of their system back during tax time. Woot! The caveat to this subsidy is that the batteries can only be charged using solar. This operating mode provides an easy-to-achieve 100% charging from solar. So does the next mode, as long as the grid-charge feature is turned off.

TIME OF USE

I describe this mode as a focused self-consumption. In this mode, the system will charge and discharge as described in the previous mode but only during specific times. This mode favors homes with ToU utility rates.

The ToU mode can use the grid to charge the batteries during off-peak times, when prices are low. Typically, ToU high-peak windows begin in the evening hours, when there is not enough sun to charge the batteries. I'm sure this is a total coincidence and an oversight by the utility provider. While it is possible only to use solar to charge the batteries, it tends to be a riskier move. If the batteries do not fill up before the high rates start, the homeowner pays the high utility prices they were trying to avoid.

Let's look at an example of the Sacramento Municipal Utility District (SMUD) ToU rate structure. It is relatively straightforward compared to other utilities.

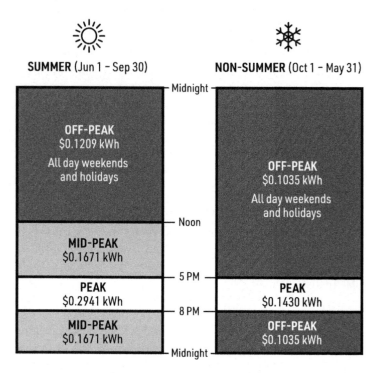

System Infrastructure Fixed Charge per Month: $21.05

This chart denotes the times of day and corresponding utility rates for two different seasons. Of note is the mid-peak time during the summer months. Don't forget the fixed charge for the privilege of being connected to an electrical provider!

"OK, so there are different windows. I get it. But what is all of this costing me?"

That depends!

If you can use less power during these high-peak times and shift high-demand activities, like washing clothes or running the dishwasher, to off-peak times, then you can save some money. You can also use a storage system to only discharge during the high-peak times and continue living your life as you wish. Many homeowners in high-peak-rate areas see a respectable ROI on their solar and storage purchase. It is conceivable to reach a five- to seven-year ROI depending on system size, including solar and storage incentives and rebates.

"Look, dude, this has gone on long enough. When will you get to the part where you tell us how big a system we need?"

Yes, I suppose it is time to dive into system sizing. Buckle up, buckaroos!

CHAPTER 9

SIZING STORAGE SYSTEMS

Sizing a solar and storage system is one of the most challenging yet rewarding experiences in this industry. As the installer, you can sleep well knowing you installed a safe, worry-free, and long-lasting system. Homeowners will enjoy energy security with a smart and clean source of renewable energy that you designed! Lastly, you can prevent unnecessary service calls caused by an improperly sized system, with fuming homeowners in the background wondering why their battery isn't working in the middle of a blackout. Not fun.

The internet has a lot of information about sizing solar and storage systems. It is tough to get into any significant detail, though. Even if you tried, there is always a goofball application someone will ask about that requires a footnote. Showing the math would probably scare people away. Instead, they make a simple tool with the unintended consequence that it can result in a poor customer experience. Oversimplicity has overtaken due diligence. Using their "crash course" approach to battery sizing might result in a crashed system. That's not a chance I would want to take.

One of the most frequently asked questions is about product selection. How does one know which storage system is the right one? These systems represent a considerable investment in technology and energy security. As usual, the internet is a hive mind of confirmation bias. The first hits of a search may not be what you are looking for but rather what everyone else found. A search for "the top 20 battery systems" is hit or miss. You are looking at lists filled with manufacturers who probably paid to have their products featured. Reputable battery systems are sometimes left out.

Most installers only use a handful of products. It keeps their ordering and stocking to a minimum by limiting part numbers, and the designers only have to remember a handful of technical specs. Also, it keeps the installation crew proficient by maintaining familiarity with the manufacturers' recommended installation practices. If the homeowner wants a specific brand, it may not be in the installer's repertoire.

The internet offers manufacturer spec sheets, top-10 lists, and unbiased, totally fair product representations disguised as customer testimonials. Well-researched homeowners may stumble across a product that interests them, but their installer does not offer it. Some companies stick to their handful of products and decline unfamiliar ones, but some dive into unknown territory. It is hard to turn down money! If I had a beer for every time someone told me, "I didn't want to install a <insert competitor>, but the homeowner wanted one," I would never need to pay for beer again.

Publications and websites that talk about sizing and designing ESSs are reluctant to recommend products, and understandably so. Products change or become obsolete all the time. Also, we wouldn't want to hurt anyone's feelings by not mentioning their product. But why go through all that trouble to try and teach someone and then leave it to them to figure out the rest? An example system goes a long way.

I will take a similar tack since there are just too many products out there. I will throw in as many examples as I can. However, as my disclaimer states, I will not be held responsible for any information in this book that becomes outdated as soon as it hits the market. Also, to reiterate, I do not endorse or recommend any of these products, and I am referencing them for educational purposes only. Unless I'm not. A quick internet search should reveal tons of competing products. So grab a beer (or other boring beverage of choice), get comfortable, and let's start slow.

Helm, all ahead one-third.

As I said in an earlier chapter, many people use their ESS for grid-up operation, primarily to offset their utility bill, but it's not necessarily sized for a significant off-grid period. These applications are the easiest to size. Some European ESS manufacturers have an order form on their website. Not a reseller, the actual manufacturer. The sales tool only asks for location, household size, solar size, and credit card number. Simplicity must be a German thing, but some US battery manufacturers also take this approach, which is risky.

"Oh, so they can do it, and it's OK, but we can't? *Huh?*"

Not exactly. The primary difference is in the application. Germans don't see the value in battery backup. Most of the battery systems sold there do not offer this operating mode.

"Why would we need backup capability if the grid is so reliable?" (actual question from an SMA engineer).

SIZING TOOLS

There are likely two different sizing tools you will use for solar and storage systems—one for the solar inverter and the other for the batteries. Both types of sizing can get tricky, and there is/ was a lot of math involved. However, some manufacturers combine the battery and array sizing estimates into one tool. Over the years, we have transitioned from manually calculating array sizes using pencil and paper to using elaborate spreadsheets and software tools.

All PV inverter manufacturers offer a sizing tool that will quickly calculate the array configuration for safe and efficient operation. In particular, one that doesn't belong to a manufacturer is on the Mayfield Renewables website.[1] Ryan Mayfield has been in the industry for many years and is a sought-after expert in understanding and interpreting NEC language. He routinely teaches daylong classes to solar installers on understanding the nuances of solar (and storage)-specific articles of the NEC.

Let's begin with a simple, low-impact online sizing tool. Several manufacturers and suppliers use a tool that asks a few questions before displaying a system recommendation for both solar and battery sizes. The first such tool is the Tesla sizing calculator.[2]

System sizing lives on the order page and states (as of this writing) the unit can power "plugs, lights, and 120-V appliances." This statement has changed over the years and appears to have settled on this conservative list of loads. Provide your address, average monthly utility bill, existing PV system size, and a $100 deposit if you want to proceed with the purchase. No specific load information is required. While I admire the ease and simplicity of this tool, it could result in a poor customer experience if the salesperson does not manage customer expectations. Let's fill it in!

Choose Your System

Tesla Recommends
4 Powerwalls
$33,000*

Edit

___

1.5 Days Backup Duration
Includes plugs, lights, 120V appliances, and some 240V appliances

Tesla Recommends
12.24 kW Solar Panels
$18,204*

ADD

$65,435 Estimated 25 year savings
Additional 10.5 days of backup
$3,000 Bundle discount

This sizing tool is quick to recommend a system size and is relatively accurate most of the time. My summertime usage is about 65 kWh per day, and I pay about $300 a month in the summer. The sales tool recommends four units, each with a 13.5 kWh usable capacity, which totals 54 kWh of capacity per day, 1,620 kWh per month. This estimate is shy of what I would need for three, possibly four, months out of the year, but it's more than enough for wintertime. So, for a grid-up application, this system would offset my summertime usage by about 83%. The battery section also states I will get 1.5 days of backup just from the batteries, assuming they have a full charge when the grid goes down.

Adding solar will extend my autonomy another 10.5 days. However, the recommended 12.24 kW solar size is another concern since I can only fit about 5 kW on my roof. I'm sure the 10.5-day estimate is full of assumptions, but if you can make it 24 hours, you probably have a perpetual backup system for at least some parts of the year. This 10.5-day estimate suggests the solar will not fully charge the batteries at some point. But if I were to change my consumption habits, I believe the system could be perpetual. In my experience, most homeowners will not change their habits. Well, eventually, they will, by force. It is not surprising how fast people change behavior when they are in the dark looking at low-battery-level warnings on their devices. See, that wasn't so hard, was it?

You may find a website that shows system packages with various loads and run times. I tend to be very critical of these over-oversimplified sales guides.

Basic package—Lights, devices, refrigerator, garage door

Deluxe package—Lights, devices, refrigerator, garage door, sump pump, wine fridge

El Presidente—Lights, devices, refrigerator, garage door, sump pump, wine fridge, three-ton air conditioner, pool pump, arc welder, glass kiln

Years ago, I made a similar sales sheet with the Big Kahuna package. Although not quite as robust as the el Presidente, I abandoned the document, much to the salespeople's dismay. It set unrealistic expectations for the long run. Homeowners won't remember the discussions concerning battery-capacity conservation, but they will remember you told them they could run an entire list of appliances and devices. Many battery manufacturer and reseller websites have removed these types of sales documents. Good for them!

SIZING IS A PROCESS

Now let's jump into sizing a bulletproof system using this three-step process:

1. Calculate the loads.
2. Match equipment to results.
3. Brief customer and manage expectations.

That's it, just like a three-point sermon, and I'll even add in the hellfire and brimstone. A load calculation is the hardest part of this process. Let's start with the data entry descriptions, which are typical for every sizing tool regardless of medium or format.

Loads—The name of the appliance, device, or whatever will be consuming battery energy

Number—The number of appliances or devices to be powered

Watts—How many watts the load requires to run

Watt-hours—The amount of energy consumed

Surge—The amount of startup power required by the load

LOADS

Generally speaking, the total number of watts tells us how big the battery inverter needs to be. For example, if the total equals 5,500 W, we need a system capable of a continuous rating of at least 5,500 W. But this rule of thumb assumes you will run all those loads at the same time. Off-gridders will argue this is how it has to be, but we aren't sizing for off-grid. Off-gridders typically know how to manage their consumption in a way that would shame their grid-tied counterparts.

This Hail Mary sizing method could result in an oversized and overpriced system for grid-tied applications, even with the contention that homeowners will use every watt you give them. A more conservative approach would be to evaluate the total watts used and talk to the homeowner about their power demands. Their "Oh my gosh, the grid is out!" mindset should not be, "Yeah, baby, let's fire up the hot tub!" If the total watts value exceeds the battery system's maximum continuous-power rating, you have reached the first decision point in your sizing. You have a few options, ranked cheapest to most expensive:

1. Reduce load.
2. Add more power.

Reducing load should always be the first go-to option. If the roof is already optimized with solar, adding more may not be an option. Adding more power means adding more battery inverters. Adding more storage systems to increase power or capacity is called stacking, clustering, daisy-chaining, connecting, and so on. The underlying architecture is that multiple units can feed the same electrical panel. Some manufacturers allow stacking, and some don't, and this is a significant feature to consider.

There are two types of loads that will strain the battery system in two different ways. Resistive loads usually contain a coil and generate heat—for example, floor heaters and electric water heaters. Technically, incandescent light bulbs are resistive, but let's focus on larger loads. Resistive loads typically strain the battery capacity, meaning the longer they stay on, the more battery they consume. Inductive loads, for our purposes, require large startup currents and typically put a strain on the battery inverter (power electronics). Inductive loads include transformers, motors, and pumps—for example, pool pumps, well pumps, sump pumps, air conditioner condenser motors. To keep the electrical engineers off my back, I will mention capacitive loads like capacitor banks and wiring, but those are irrelevant for most residential applications.

RUN TIME

"Run time" refers to how many hours per day the load will run. Some transient loads, like microwaves, garage-door openers, and toaster ovens, have daily run times in fractions of an hour.

Some loads, like refrigerators, pool pumps, and air conditioners, do not run all day. They have run cycles, meaning they run for a particular time, turn off, and then start up after a downtime period. On average, a refrigerator will run about six hours per day. You can hear the compressor start up after you leave the door open while trying to find the banana Moon Pie you hid the night before. An air conditioner will run about six to eight hours a day, as will a pool pump. Take an educated guess at how long you will leave the lights on, watch TV, or run the microwave for all other loads. In a post-COVID world, sizing rules of thumb may need to be tweaked a bit since more of us work from home.

Things like garage doors and microwaves put such a small strain on the battery system I don't even consider them unless I want to fill the list to make it look impressive to the homeowner. A single 2 kWh battery can cycle a garage door about 20,000 times. I also (somewhat) debate microwave use since Jimmy Dean sausage biscuits only take about two minutes to cook. Still, it would be best to reduce that transient variable as much as possible if you have teenagers.

WATT-HOURS

A watt-hour is the calculated value of watts multiplied by run time. The total watt-hours value provides the initial battery bank capacity. Although the homeowner may not turn every load on that list on at the same time, they still use an amount of energy that is consistent throughout the days, weeks, and months, depending on the season. For example, your utility bill parses out energy consumption in kilowatt-hours, but unless you are in a special area with demand charges, you will not see a power rating or value in watts. So, in the summertime, you may not have every load on in your house at the same time, but your utility bill states an average daily consumption of 25 kWh. Your daily consumption is true. This is the value the storage system needs to match more closely than the total watts. Unless you have a customer who needs large amounts of power.

SURGE

This variable is arguably the most important, but often overlooked, value in most load calculations. The surge rating is the amount of current it takes to start a load. Some manufacturers will state a maximum output value for a specific amount of time, usually in milliseconds or seconds. Surge ratings of large loads put a tremendous strain on the battery inverter's ability to overcome the device's locked rotor amp rating. The sum of this column would represent the worst-case scenario if all large loads were to turn on simultaneously. Surge is also called "startup" or "inrush current."

If the total surge is less than the battery-system surge rating, you are good to go.

If the total surge is greater than the battery-system surge rating, you have reached a decision point:

1. Reduce load.
2. Add more surge capacity.

Adding more surge capacity means stacking more battery inverters.

"So, where am I supposed to find this surge rating on these loads? Does everything have one?"

That depends!

"For which question?"

Both!

"You suck."

The best way to ensure the correct surge rating on the load sheet is to either measure it or find it on the device nameplate. Measuring a surge takes some sophisticated equipment since most multimeters cannot catch the spike. Reading the nameplate value is the next-best thing. Most battery systems can start a refrigerator or freezer compressor, so I wouldn't worry about pulling these appliances out from the wall and digging around for the compressor rating. Energy Star–compliant refrigerators will take about 8 to12 A to start and consume 1.5 to 2 kWh per day. Pumps, motors, and compressors are a different story. The largest load people usually want to back up is the central air conditioner. For the record:

I do not condone the placement of central air conditioners on the load side of a battery system. They will chew up battery capacity faster than a dog gnawing on a ham bone. Relatable example, I'm sure.

However, for grid-up applications, like ToU or self-consumption, the air-conditioning unit is a widespread load. Therefore, the following information still applies to anyone sizing for these operating modes since the battery will switch to backup if the grid goes down.

"Yeah? Tell that to someone in Phoenix, where it's 120 degrees in the shade. In May."

Every location has its environmental considerations. Instead of backing up the central air conditioner, it might be wise to have mini-splits installed in strategic rooms of the house, like the master bedroom, for example. A mini-split will only use 5 or 6 A and will not strain the batteries as much. Service logs are littered with cases containing the search phrases "battery shut down," "deep discharge," "air conditioner," and "upset homeowner."

Air conditioner surge values, listed as "locked rotor amps" (LRA), are hard to find in manufacturers' manuals. The values are so close between makes and models it is tempting to use a rule of thumb like "a three-ton air conditioner has an average LRA of 77 A." For example, a Ruud three-ton air conditioner model WA1636 has an LRA of around 84 A.[3] A Trane 2014 model of the same size will have an LRA of 79 A. However, a Lennox three-ton air conditioner will show a lower LRA rating of 64 A.

The best way to find the rating is to look at the air conditioner unit compressor hidden in the condenser unit. The condenser unit has a big, loud fan. You will need to isolate power to the unit, remove the condenser service cover, and then find the compressor label. Look for a rating that says LRA—this value indicates how much current it takes to get the rotor to turn over from a stationary position. Think of your car engine when you turn (or push) the key. The tachometer revs up to 5,000 then slows to about 1,000. Of course, if you own an electric vehicle, this is not a relatable description.

As an example, let's look at a three-ton compressor with an LRA of 76 A.

The next step is to bounce this value off the battery-system spec sheet. The designer is speccing a Powerwall, and the max peak current (10 seconds) is 7.2 kVA (kilovolt-amperes; 30 A at 240 V). As we can see, like most ESS products, the air conditioner compressor rating is higher than a single Powerwall. Not a problem since these batteries can stack. However, someone should have a conversation with the homeowner about using these large devices when the grid goes down and alternative forms of climate control.

Battery inverter surge ratings are a spectrum:

Electriq PowerPod = 31.2 A	Sol-Ark 12K = 83 A
Generac PWRcell = 41.6 A	SolarEdge StorEdge 7600A = 34.7 A
LG Chem RESU10 = 29 A	sonnen's sonnenCore = 36 A
OutBack Power SkyBox = 24 A	SunPower SunVault = No online specs available
Schneider Conext XW Pro = 52 A	SMA Sunny Boy Storage = 38.75 A

These surge values are the converted maximum output power rating gathered from the spec sheets. See, that power formula is a handy one to have around! So far, we aren't doing so well with products that can handle our three-ton air conditioner's surge. Only the Solar-Ark, or an older Sunny Island if you could find one, has a surge rating that exceeds our requirement. However, most of these models can stack, which means we can eventually reach our 76 A requirement.

"But what if two of those battery inverters only get us close to the LRA? Is there no room for slop with these surges?"

You can do three things to reach the rated surge (LRA) of a pump or motor:

1. Add more battery inverters.
2. Use a soft-start device.
3. Upgrade the air conditioner to a more efficient model (and keep your fingers crossed).

A soft-start device reduces the amount of current draw using capacitors or solid-state components. Depending on the rating, it could reduce our three-ton air conditioner's draw to the range of our battery inverter. Considering that a soft-start "kit" will run about $200 to $400, it is usually cheaper than buying another battery inverter. As an example, and not a product endorsement, the EasyStart 368 soft starter claims it will reduce the LRA by up to 75%. Our battery inverter would only have to supply 25% of the 76 A requirement, a measly 19 A. One caveat to these soft starters, regardless of make or model, is that each large load will usually require one.

All these techniques, as mentioned above, apply to other large loads as well. Pool pumps, basement sump pumps, heat pumps, air compressors, arc welders, elevators, car lifts, and so on, will have labels with their surge requirements. However, some pumps do not explicitly reveal their surge ratings. We must decipher this puzzle using their code rating. They don't make it easy, do they?

On every AC motor label, there is a section that says "code." The motor is what drives the pump's compressor. The label doesn't tell you what this code is for or why you should look at it in the first place. Here is an image of my parent's pool pump. It is a Hayward 1 HP pump connected to a 20 A breaker in the barn.

Notice there is no LRA rating printed on this label, but the National Electrical Manufacturers Association (NEMA) Code will give us the same information.

NEMA CODE LETTER FOR AC MOTORS

The following table provides the kilo-volt-ampere per horsepower range for each NEMA designation.[4]

LETTER DESIGNATION	KVA PER HORSEPOWER
A	0.0–3.15
B	3.15–3.55
C	3.55–4.0
D	4.0–4.5
E	4.5–5.0
F	5.0–5.6
G	5.6–6.3
H	6.3–7.1
J	7.1–8.0
K	8.0–9.0
L	9.0–10.0
M	10.0–11.2
N	11.2–12.5
P	12.5 -14.0
R	14.0–16.0
S	16.0–18.0
T	18.0–20.0
U	20.0–22.4
V	22.4 +

Now that we have the NEMA Code decoder ring, we just need to locate the pump's code value. The Hayward 1 HP pump has a code of J. Skimming down the kVA column, J equals a kVA startup rating of 7.1 to 8.0 A. This startup would be no problem for any of the previously mentioned battery inverters—or one

from any other manufacturer. An 8 A startup is noob-level weak
sauce for most battery systems.

POINTS TO PONDER

What taxes the battery capacity more: many small loads running
simultaneously or a few large loads that only run for a few hours
a day?

The answer: it depends! And it always will.

The two most common questions people ask me when trying to
calculate a battery bank size are:

1. "I have a customer who uses 3,000 kWh a month. How big a
   battery does he need?"
2. "I have a customer with a 2,500-square-foot house. How big
   a battery does she need?"

My reply to the second question is always the same. "Oh, 2,500
square feet? What color is the house?" My sarcastic point being
you cannot calculate consumption based on square footage. I
have seen a 4,000-square-foot house that was more energy efficient than a 1,700-square-foot house. Generator manufacturers
would disagree, but as long as there is fuel, a generator is good
to go.

The first question has a caveat. If a homeowner only wants to
reduce their energy usage from the grid (ToU or self-consumption)
and does not care about backup, monthly use is probably sufficient. However, as I said, these systems will always shift into a
backup mode when the grid goes down. And when it shifts, the
battery will be feeding for a grid-up application.

Ideally, the PLP should contain only *essential* circuits and receptacles. Essential loads include the refrigerator, freezer, security system, garage door, lighting, and a few outlets for devices. Others want pool pumps, central air conditioners, and large electric appliances like water heaters, glass kilns, and stoves. Generally speaking, these large loads are energy hogs and will deplete a respectably sized battery bank in record time. However, managing customer expectations goes a long way.

I took a very memorable service call back in August of 2020. Unfortunately, monitoring screenshots do not lend well to print. Pity since a picture is worth a thousand words for service calls. This frustrated installer was on-site and called me for some troubleshooting advice. The home had lost power at around seven o'clock the previous night, and the system was in low-battery mode due to a deep discharge. I pulled up the unit number on our back-end portal and saw yet another familiar story. The unit switched to backup mode when it detected the power outage. It covered the small nighttime loads with no issues until about five o'clock the next morning. The battery discharge jumped from a baseload of fewer than 500 W to a staggering 5,000 W! The SoC quickly fell from 60% to 5% in about an hour. This customer had no power when they woke up and no sun to help.

Looking at previous days, I asked what device turns on every day at around 5:00 a.m. I heard the conversation in the background after the installer pulled the phone away from his face.

"You have a hot tub on there?" rang out loud and clear in my earpiece.

Maybe I am old-fashioned, but I don't think a hot tub is an essential load. This one put too much strain on the battery, leaving the family stranded. As I said, managing customer expectations is the most critical thing you can do to keep everyone happy.

"So I am supposed to ask the homeowner exactly what loads they want to run and for exactly how long after running around looking at the appliance and pump labels? That seems dumb. Isn't there an easier way?"

Of course there are easier ways, and I will discuss some sizing methods later. Load lists are a good guideline and place to start. They do not account for the person who leaves the device on for 30 minutes longer than what is listed. They do not account for that brand-new swimming pool bought with the annual bonus five years down the road. While utility bills provide a glimpse into our consumption habits, they do not show the entire picture. But there are devices that will.

There is a growing tendency among ESS installers to nail down load and consumption habits using specialized, commercially available monitoring equipment. These systems use clamp meters to measure specific circuits to provide a load profile. Some of these monitoring systems are sophisticated enough to recognize particular loads at given times—such as refrigerator, dishwasher, lighting, and microwave. Neat! These systems always have a visualization tool to see at what times consumption is highest and which loads are causing the energy or power spikes. Installation companies that use this load-calculation method report higher customer satisfaction with the installed storage system since the load profiles are nailed down almost to the watt-hour. The process goes something like this. During a presale appointment, the salesperson (usually) discusses initial system sizing, needs, and so on. The salesperson will also inform the homeowners of a future visit to install the monitoring device. Someone shows up after the presale meeting to install the monitoring kit. They return a month later and analyze the data. Then the salesperson contacts the homeowner with any system-size adjustments—bing, bang, boom. Done!

There can be challenges with this method. First, salespeople show a high reluctance to use this process. Any guess why?

It slows the sales process. I can get around this with a simple conversation that leads the customer to their conclusions about due diligence and budgetary constraints. "Although we have talked for a while about your loads and solar, this estimated system size could change. We will install a monitoring device so we nail down your consumption patterns and behaviors and then return with a more precise system size. This new size could be larger or smaller than this estimate." To ease the risk of losing a customer in the meantime, the salesperson can do two things. First, take a deposit based on some appropriate "consultation fee" to make the impending sale more real to the homeowners. Then, if there is any pushback, vilify your approach. "I know a month seems like a long time to wait, but we prefer this method since we can get a better idea of what you need. We might be able to save you money on a smaller system. If you prefer, we can even leave the monitoring devices connected as part of the sale. Besides, wouldn't you rather have this kind of due diligence than someone who just comes along and gathers just enough data to spit out a system that may not even be what you really need?"

There is product selection galore for these monitoring devices based on how much detail you need and how much you want to pay. A quick internet search for "home energy meter" will reveal the big names first. Smappee, Sense, eGauge, and CURB usually dominate the first page, but scrolling past page 1 (gasp!)

shows more options. Considering the cost of whichever product you use, leaving the monitoring system behind may be worth it, assuming you would have to roll another truck to go back and disassemble everything.

That covers the sizing-tool data interpretation. Now let's do one for real and take a look at compatible systems.

CHAPTER 10

CALCULATING LOADS

There are as many sizing tools as people who make them, but the tools all use the same math—although some use more than others. Luckily, the beginning process is always the same. Since backup is the most challenging mode of operation, let's calculate those loads using this checklist:

1. List all loads powered by the battery system.
2. List the wattage for each load.
3. List the amount of time each load will run per day.
4. Multiply watts by run time to get watt-hours.
5. List the surge rating for each load, if applicable.

Here is an example of a sizing tool in boring table form. This form is the backbone of all sizing tools regardless of delivery medium.

LOAD	QUANTITY (1)	WATTS (2)	HOURS PER DAY (3)	TOTAL WATTS (W) (4)	TOTAL WATT-HOURS (WH)	SURGE AMPS (A)
LED lights	10	15	6	150	900	0
Fridge	1	300	6	300	1,800	12
Phone	3	10	1	30	30	0
TV	2	200	6	400	2,400	0
Coffee maker	1	1,000	0.2	1,000	200	0
Microwave	1	1,000	0.2	1,000	200	0
Router	1	40	24	40	960	0
Security	1	50	24	50	1,200	0
TOTAL				**2,970**	**7,690**	**12**

To find total watts, multiply watts (2) by the number of loads (1).

To find total watt-hours, multiply total watts (4) by hours per day (3).

Total power required if all loads are on simultaneously: 2,970 W.

Total watt-hours if all loads (and not one load extra) run for the *exact* number of hours estimated per day and not a single minute more: 7,690.

The bolded totals reveal key information about our system requirements:

Total watts = Size of the battery inverter output

Total watt-hours = Size of the battery bank

Total surge = Maximum power or surge rating of the battery inverter

"Oh, so we are done with this, then? Looks like we need a 7.69 kWh battery."

Not exactly. This is called rough sizing, and it only gets bigger from here. If the rough-sized battery bank size is scary, the finished product will be horrifying. We'll keep this initial sizing on the back burner for now.

ONLINE CALCULATORS

Load-calculation methods vary from designer to designer, partly in the number of extra variables they include in their calculations. Although I still know a few people who do these calculations longhand using a pencil and a sizing table like the one earlier, most have moved on to electronic versions. Sizing tools abound on the interwebs. An internet search for "battery sizing calculator" finds the Trojan Battery sizing tool at the top of the search results, probably because they had the foresight to secure that URL. OK, let's start there.[1]

Trojan has a brilliant and easy-to-use sizing tool that will, of course, recommend an appropriate Trojan battery for your project. Although lead-acid battery sizing is beyond this book's scope, we can cheat and lift some excellent baseline information about our system using this site. This tool does have a lithium option. Let's fill in the first section (Step 1) for lithium-ion battery application.

RE STEP 1

Please Select Your System Design Parameters

Choose system design battery voltage (12V, 24V, or 48V)

| 48 V | ∨ |

Choose type of your PV system

| Grid-Tie | ∨ |

Choose desired battery depth-of-discharge (DOD)
DOD describes how much of the total amp-hour capacity is used during a discharge cycle and is expressed as a percentage of its rated capacity (Select 40% as the DOD if you wish to take no more than 40Ah from a battery rated at 100Ah)

| 80% | ∨ |

Type of Battery

| Lithium-Ion | ∨ |

Days of Autonomy
This is the number of days the battery has to power the designated loads (Select 2 days if you want to power your loads for the desired durations over 2 days)

| 2 | ∨ |

Now, it's time for the fun part—Step 2, entering loads! The Trojan tool offers a link to an Energy.gov site with common household loads to try and ease the pain of this lengthy part of the process.[2] However, the Energy.gov site asks the user to enter the same information the Trojan tool asks for. How many people know how long their refrigerator compressor runs? Or how many watts it takes to run? The Energy Star website offers a Flip Your Fridge tool that will give you energy savings and annual kWh usage, but you will have to do some math to make it fit the tool.[3]

For example, the Energy Star tool states my 24.5-cubic-foot or more refrigerator will cost about $61 a year and consume 508 kWh annually. Fine, but the battery sizing tools ask for run time and power requirements.

508 kWh per year / 365 days = 1.4 kWh per day

This value is close to the industry-accepted consumption of 1.5 to 2 kWh per day for a refrigerator. Unfortunately, most tools will not accept the kWh value. They require the device's wattage rating and the amount of time it will run per day to calculate the kWh value. Fun stuff, isn't it? The average run time for a refrigerator is about six hours per day. Punching this into the power formula will tell us the amount of wattage this device uses.

Watt-hours = Watts × time, so let's solve for x

2,000 = x × 6

2,000 / 6 = x

x = 333 W

So, it is safe to say an entry of 333 W with a run time of six hours is good enough for a refrigerator (rounding to 400 W probably wouldn't hurt). But not every appliance and device has an Energy Star rating or a printed wattage on the sticker. For example, my printer plugs into a 120 V outlet, but that doesn't tell me much. The sticker does not provide this information either. The best thing to do is use rounded numbers found with another internet search. "Home appliances wattage lists" abound and provide generally accepted values for most everything in the home. A Kill A Watt meter is also a great tool that reveals load data.

"Well, sheez, why didn't you just start there?"

Because most of these lists only provide daily kilowatt-hour values. Math is still involved. The shortest time they will allow is one hour, which is problematic for devices like microwaves, toasters, and garage doors that run for fractions of an hour each day. I filled in the loads for Step 2 using the previous chart.

Notice how the Trojan approach allows decimals and includes a 15% inverter efficiency loss in the calculation. I love this conservative approach. The next step presents the results. Drum roll, please!

click

"We are unable to recommend a Trojan solution based on your current system design inputs. Please review the values you entered for system parameters in Step 1 and load estimates in Step 2 to be sure they are accurate."

"Wait, what? After all that, they can't even recommend a battery?"

Not exactly. Looking back at Step 1 and the lithium-ion battery selection, it appears this is throwing off the calculation since the manufacturer does not offer a compatible battery for the final numbers. Changing the battery-type requirement to "any" presents five lead-acid batteries that will meet our needs. However, although Trojan cannot provide a lithium-ion solution, they still deliver the overall storage requirements in another box that can help us with our product selection, albeit with a different manufacturer.

System Loads and Battery Capacity Requirements

Values below will change as you enter system parameters and load estimates.

Battery watt-hours per day for AC loads (including 15% AC inverter loss)	9753	Wh/day
Battery watt-hours per day for DC loads	0	Wh/day
Total battery watt-hours per day (assuming 97% wiring and distribution efficiency)	10055	Wh/day
Avg daily battery amp-hours needed (with 48V battery system)	209	Amp-hours (@48V)
Required system capacity* (based on desired 80% DOD) to achieve 2 days of Autonomy.	524	Amp-Hours (@48V)

This project requires a battery bank capacity of at least 10,055 Wh, and this is enough information to start looking at compatible systems. Now let's switch gears and look at a few manufacturer online sizing tools.

SunPower has an online tool for its storage system that uses a dissected 3D home that lights up the house loads when selected.[4] It asks for home square footage and solar system size and does not allow for more loads. There is a heavy disclaimer on the page, and like most tools of this type, there is a form to fill out for a sales callback. I hope these callbacks include managing expectations.

Let's take a look at another online sizing tool that blends the quick sizing approach with a touch of thoroughness. The Enphase System Estimator begins its sizing tool by asking for your address and either (1) monthly utility bill total or (2) square footage of your home. Are you detecting a pattern?

⊖ ENPHASE | System Estimator

| I'm new to Enphase | I'm an existing customer |

● Enter your requirements

My Zip Code/Address

My Home Size ▾   210(sq. ft

Estimated energy consumption 30 kWh/day ✏️

✅ Solar   ✅ Storage

In an outage, I need a backup of   − 12 hrs +

My home appliances	I own	Use during outage
Air conditioner	✅	⬤
Pool pump	○	⬤
Electric vehicle	○	⬤

Backup ACs and Pumps with Enphase Power Start™ technology.

Show My System Estimate

● Refine your estimate

● Download your summary

● Connect to Enphase

It is a reasonably accurate tool and differs from the previous
Tesla tool in that you can fine-tune the load profile. Additionally,
there are sliders to adjust PV system size and battery capacity
manually. Lucky for me, since I am not able to fit the recommended 56 "panels" on my roof. I slid the bar down to 16, which

is what my roof can accommodate. Before I even reached the target number, red text boxes appeared to tell me that my new solar array size would not be sufficient to recharge my batteries. Attention-grabbing text is an excellent tripwire since it now forces an end-user conversation about loads and consumption.

The last online tool is the Big Kahuna of all free online tools. It goes into the most depth and has the most deliverables of any manufacturer. SMA Sunny Design requires registration,[5] but you can design a project to electrify an entire off-grid village if you want. This tool has matured over the years, but it is not for the faint of heart. It is the equivalent of a high-level, fully geared dungeon mini-boss. Sunny Design focuses on load profiles and then recommends the appropriate PV and battery configurations. There is even a section for generators. I have used this tool for several off-grid projects using lead-acid batteries with great success.

SIZING SPREADSHEETS

Also available using an internet search engine, spreadsheets will ask for the same information as the online tools. Search for "battery sizing spreadsheet," and in half a second, you will have well over half a million hits for Excel, Word, PDF, and other formats.

Let's finish this up using some age-old rules of thumb that still apply. All current battery bank sizing methods are based (sometimes loosely) on how off-gridders have been sizing battery banks for decades. Since we already calculated total watts and total watt-hours, we simply plug them into the master formula. I tried to squeeze a *Tron* reference in there, but I couldn't make it work.

SIZING FORMULA (LEAD-ACID)

Battery size =
(Daily amp hours / inverter efficiency) × battery multiplier
× days of autonomy / discharge limit

You can find the battery multiplier on the manufacturer's spec sheet. I will use 80 degrees Fahrenheit, which has a multiplier of 1. Lead-acid batteries typically only warrant a 50% DoD. Off-gridders use a three-day autonomy factor in case there are some terrible solar-production days.

An estimate of a 10 kWh daily load requirement for a lead-acid battery bank would look like this after filling in the formula above:

$(10,000 / 0.9) \times 1 \times 3 / 0.5$

$11,111 \times 3 / 0.5$

$33,333 / 0.5$

$= 66,666$ Wh or 66.6 kWh

You can see how the battery bank exploded from our initial estimate of only 10 kWh, mainly from the lead-acid 50% DoD limitation. The three-day autonomy is also a substantial addition. Although a very common calculation for off-grid, it is subjective for grid-tied applications. Most salespeople will not use this autonomy add-on since it will probably price them right out of a sale. However, the California wildfires and mandated utility shutdowns have shown power outages can easily extend to 10 or 14 days. A PV array covered in soot and ash will not produce as much as a clean array, which is what the designer based the sizing estimates on in the first place. Instead of sizing for an enormous battery bank, a small generator would suffice along with cutting down on energy usage.

"OK, we get it already with the off-grid crap. What do the rest of us do?"

Let's tweak the formula for lithium-ion applications.

Battery size = (Daily Wh / inverter efficiency) / discharge limit

One could argue we could disregard the discharge requirement since most lithium-ion battery systems state a 100% usable capacity. But not all of them do, so we'll leave it in. If a manufacturer says 90% usable capacity, the discharge limit would be 0.9 since we use DoD.

Battery size = (10,000 / 0.95) / 0.9 = 11,695 Wh

First, I would round up to 12 kWh. Next, although a two- or three-day autonomy might price the system out of the homeowner's budget, you should still bring it up. They need to understand they do not have an infinite source of power when the grid goes down. Some might, but those are usually smaller systems with small loads and a beefy solar system. I will address adding solar to the calculations later.

Once we calculate the final battery size requirement, we need to compare the values with the storage system manufacturer's spec sheets.

1. IS THE TOTAL CALCULATED WATTS VALUE GREATER THAN THE BATTERY INVERTER CONTINUOUS OUTPUT?

Yes—Get a bigger inverter (proceed to next decision point).

No—Battery inverter size is good.

2. IS PURCHASING MORE BATTERY INVERTERS AN OPTION?

Yes—Good to go.

No—Reduce load.

3. IS THE TOTAL WATT-HOURS LESS THAN THE BATTERY-SYSTEM CAPACITY?

Yes—Good to go.

No—Get more battery capacity or reduce the load.

This "yes" part in Step 3 can get tricky. For example, suppose you consider battery-only systems like Blue Ion,[6] LG Chem,[7] or Tesla Powerwall.[8] This step is complete since those battery systems can stack. However, suppose you are considering a manufacturer that cannot stack. In that case, the options may be limited if your required capacity exceeds what one unit can provide.

For example, the Generac PWRcell has four sizes based on capacity: 9, 12, 15, and 18 kWh.[9] You can connect two PWRcell batteries to one of its inverters to increase capacity. If the calculated battery size is 30 kWh, you could easily match PWRcell's battery choices to your calculation. Two PWRcell 15 batteries will do the job. You could even use two PWRcell 17s to add a little more capacity for future use. However, if you need more than 36 kWh, you will need to use more PWRcell inverters to accept that extra capacity. There are always workarounds, but that adds cost and complexity.

Adding 20% to 30% of battery capacity to the initial battery calculation to account for future growth is a good idea. However, this might prove tricky depending on the ESS architecture. It would be wise to talk to the homeowners during the presale meeting to see if they plan on adding any large loads. Which they will.

GREG'S AXIOM #4

Homeowners will use every watt-hour you give them. Oversizing the system to account for future growth is probably a good idea.

I will never forget a service call about a guy who drained his lead-acid, 1,200 Ah (57.6 kWh) battery bank in record time. After asking the right questions, the tech finally found the culprit. The homeowner thought it would be a good idea to fire up a 20 kW glass kiln during a power outage. After asking a few more canvassing questions, the tech learned the glass kiln was not on the original load list. Shocker.

Although this is an extreme example, many smaller loads added to a smaller-sized battery bank will have the same effect. Please take my advice and upsize as much as possible. If the initial product size is a 10 kWh model, round up to the next available size and then add another 20%. You will thank yourself later.

BATTERY INVERTER SIZING

Battery inverter sizing is straightforward. As I alluded earlier, take the total load watts and find a comparable system with at least that much output. For example, the total wattage from our load worksheet is 2,970 W. We would need to find a system with at least that much continuous output. The good news is any ESS on the market could supply this modest system.

Larger output requirements might take more than one battery inverter depending on the battery system's architecture. However, this is still off-grid thinking and assumes all the loads on that list will be on simultaneously. While possible, it is not likely. Using this method would result in a system that is much larger than necessary. If they do not exceed the inverter rating for too long, we can focus on the battery-capacity requirements. If a customer has a large power requirement, it would be wise to go over it in more detail.

CHAPTER 11

SIZING FOR OPERATING MODE

This chapter will analyze a cross-section of offerings current as of this writing. However, the mechanics involved in comparing spec sheets will be a transferable skill if you stumble across a product I do not cover. Leveraging the how-to in previous chapters, I will skip the detailed calculations for battery inverter, PV inverter, and battery bank sizing. I will focus only on battery selection. I will cover the sizing process for the common modes of operation: self-consumption, ToU, and backup.

Although you may size and install an ESS strictly for grid-up operation (self-consumption or ToU), the reality is that the house will lose power.

"Oh, we are just using this fancy battery to save money on our high ToU rates. The power doesn't go out very often, so we will just rough it when it does."

—ARIZONA PUBLIC SERVICE CUSTOMER

"Hey, why did my battery only last two and a half hours? Now I'll be in the dark all night!"

—SAME ARIZONA PUBLIC SERVICE CUSTOMER SIX MONTHS LATER

The harsh reality is that the overwhelming majority of storage installations must consider the backup capability even if the original design is for a grid-up application. But let's start with our project details and the backup method first.

PROJECT DETAILS

Total watts—6,000

Total watt-hours—15,000 (one day)

Total surge—9,600 W (40 A)

PV array—6,500 W

PV inverter—6,000 W

Average PV production—8,400 W

Location ZIP code: 01007 (Belchertown, Massachusetts. No particular reason for this location other than the ZIP code reminds me of James Bond and the town name reminds me of *Bob's Burgers*. Win-win!)

BACKUP APPLICATION

We will use Blue Planet Energy's battery for this first design, building on the previous chapter's lessons. Marketing materials (carefully and ambiguously) state this battery integrates with your "preferred inverter brand." Searching for inverter compatibility will prove to be a frustrating exercise. I can see this battery paired with the OutBack Power Radian inverter in the wild, and after digging around, I found an OutBack Power application note confirmation. Marketeers are reluctant to state specific make and model compatibility since it could change over time. Instead of updating marketing materials, they pass this pain in the ass on to us! I did manage to find a solar- and battery-supply site that offered the Blue Planet Energy batteries. The site states that there are several compatible battery inverter brands, including OutBack Power,[1] Schneider Electric,[2] and Sol-Ark.[3] OK then, let's go with the Sol-Ark solution.

Pro tip: always check product compatibility!

The Sol-Ark datasheet is heavy on marketing and promoting their solution's high efficiency. It is the only battery inverter on the market that overtly states its resistance to electromagnetic pulses (EMPs) and solar flares, appealing to the prepper in all of us. The Sol-Ark 8K inverter has a continuous output of 8,000 W and a whopping 20 kVA (83.33 A) of surge at 240 V. Although the Sol-Ark can accept up to 11 kW of solar for a DC-coupled system, the AC-coupled PV limitation is 7,000 W, typical of most battery inverters. The Sol-Ark easily meets our AC requirements for power and surge and the 6,000 W PV inverter size.

"But what if the PV inverter were larger than 7,000 W? There are a lot of 10 kW PV systems out there."

This is true. For many areas, 10 kW was the sweet spot for solar. Many incentives and rebates allowed systems up to this size. Battery inverter manufacturers have recommendations for meeting max PV inverter design criteria ranging from "You can't" to more sophisticated solutions using relays to clip strings or branch circuits. Well beyond the scope of this book, but a perfect question for the manufacturer's application engineering team. The Sol-Ark overcomes this limitation by offering a terminal block for a PV array inside its chassis. It is both an AC- and DC-coupled product.

The datasheet states it is compatible with lithium-ion batteries. The datasheet also stipulates a 48 V battery bank voltage, which immediately excludes the LG Chem and Tesla solutions since they use high-voltage DC, which is just a caveat since we already knew we wanted to use the Blue Planet Energy battery. Sol-Ark can accommodate battery bank sizes from 90 to 2,000 Ah. Some quick math converts this amp-hour range to 4.32 to 96 kWh. Our 15 kWh requirement is definitely in range, but we need to check the battery datasheet to find out how many batteries we will need.

Blue Planet Energy utilizes a 2 kWh battery, and the super cool, blue-LED-lit battery cabinet can hold 8, 12, or 16 kWh of capacity. These cabinets are modular and can stack up to 448 kWh. That's a lot of cabinets! The battery-cabinet measurements (rounded) are 42 by 24 by 24 inches; 448 kWh would definitely take up some real estate. We can quickly achieve the 15 kWh project requirement with the 16 kWh solution. Bumping up to the next available size is always a good idea!

Unforeseen circumstances, like a change in customer budget or product availability, can render a design useless in a matter of minutes. Luckily, most products are interchangeable. Swap out the existing specified product with a similarly sized product and keep going. This approach goes for battery inverters, PV inverters, solar modules, and anything else in the ESS.

Leaving a lithium-ion storage system in backup mode seems counterintuitive to me. These systems have excellent warranties, high cycle counts, and deep DoD (90% plus). You are throwing money away by waiting around for the grid to fail before taking advantage of this battery. You should put the system in a grid-up mode (ToU is a good choice) to save a little on your monthly bills!

If placed between the utility meter and the main panel, that ESS is now at the mercy of every home load. Not a big deal until the grid goes down. However, some AC panels (like combo panels) cannot accommodate this electrical option. The utility meter connects directly to the main lugs using bus bars. You can't cut those bus bars real quick and put in an ESS. Meter main combos either require a tap for a whole-home application or a PLP with all circuits shifted over. I am a big fan of PLPs.

Pro: a PLP configuration is less risky since the installer can control the loads that consume energy during a blackout.

Con: you must account for each circuit on that panel during the sizing exercise.

The customer's utility bills will not help now unless you shift all house circuits to the sub-panel. The utility bill shows total consumption, but if the design only includes the kitchen, master bedroom, garage, and security lights, those loads must be calculated separately, as we did in the previous chapter.

Sizing for a backup application is like sizing for self-consumption or ToU depending on how much of the house will be covered by the ESS. However, a bit more customer-expectation management is in order. The grid is not there to act as a backup generator to charge the batteries or cover large surge requirements. This responsibility rests solely on the ESS. The mechanics are the same, and the formulas used for these grid-up scenarios will dial in the appropriate system requirements. Product selection hinges on these requirements.

LOAD CALCULATIONS VERSUS MANUFACTURER'S SHORTCUTS

Many ESS manufacturers have enticing lists of things you will run with their products when the grid goes down. Treat these lists as an initial, rudimentary, sales-only approach to customer-expectation management and product selection. Beware, homeowners eventually use them as gospel. Regardless of how many times you paint the picture, there will be homeowners who forget all of it except that cute marketing piece that showed a hot tub, air conditioner, and pool pump.

"Hey, <Brand X> said they could power my whole house, including my three-ton air conditioner, with only two of their units!"

Well, sure, any of these systems can. The bad news: it may only be for 30 minutes. But they can do it! Proceed at your own risk and with great caution.

BATTERY BACKUP CONSIDERATIONS

You must break down the loads and look at the surge requirements. When the grid is out, loads are reliant solely on the ESS. Note the LRA ratings of pumps and motors.

Scrutinize the total required power. How many of the loads could turn on at the same time? If the amount is greater than the continuous output of a single ESS, you will need to double (or triple) up. But if only one-half or three-quarters of the load will be on at any given time, you could save some money on a smaller system. Sizing to the worst-case zombie-apocalypse or doomsday scenario is usually not necessary.

Analyze the time of year the grid power is likely to go out. Recommend alternative forms of climate control, like mini-split air conditioners or kerosene heaters. Preserving battery capacity during times of extended outages is crucial.

Consider a generator, especially in areas with low sun hours. I don't know how many times customers have called me about a homeowner who drained their battery down because of an extended outage during a week or two of bad weather. Hurricanes and other major weather events could keep the power down for much longer. The solar just can't keep up with consumption and charge the batteries in these situations.

It is a good rule of thumb to load out the ESS with as much solar as it can handle. For example, if the calculated PV size is 5,000 W but the battery inverter can handle up to 8,000 W, try to get as close as possible. I don't understand the people who don't want to send a single watt-hour to the grid, hoping they are sticking it to the man in some abstract way. Like owning a gun, it's better to have it and not need it than to need it and not have it. (Cue controversial hate mail in 3...2...1.)

Use load controllers or smart panels to control consumption. These high-tech devices allow remote breaker control and integrate with ESSs. Imagine using intelligent or smart breakers an ESS can remotely control! This new tech makes a whole-home backup a reality without the worry of draining the battery. For example, when the grid goes down, these systems automatically turn off breakers that feed large loads—such as hot tubs and central air conditioning. When the grid comes back, the breakers automatically return to their original positions. Neat!

Leviton,[4] Lumin,[5] and Span,[6] among others, offer this "smart home" capability. They aren't cheap, but the convenience and energy security are worth it for some people. Who has time to walk over to the west side of a 10,000-square-foot home to turn off the lights? A more relatable function is to turn off nonessential loads when the grid goes down. This high-tech load shedding will save battery capacity and ease the sizing requirements for these projects. As with most things, technology advances far faster than we can accommodate sometimes.

"Hey, I can do all that smart home stuff with my Alexa and smart dimmers, shades, bulbs, and switches. Why do I need all of this extra stuff?"

Well, my friend, what you have is a *connected* home. Sure, you can program lights to turn off at certain times of the day and tell your home-controller device to draw the shades. But you have to ask it to, and it may not happen automatically when the grid goes out.

OK, here is another exercise using the Generac PWRcell. This solution offers two models based on AC power ratings—the X7602 rated at 7,600 W and the X11402 rated at 11.4 kW. Both models satisfy the 6,000 W power requirement for our project. An extra battery inverter is not required since it is internal to the PWRcell. A Generac generator seamlessly pairs with this solution. The PWRcell's surge, 50 A for three seconds, meets the 40 A total surge requirement. Note that this will not be enough to start a central air conditioner. However, it will be plenty of power for small pumps and motors.

This system will need 15 kWh of capacity. It took some digging around, but I finally got the battery datasheet from a buddy who works at Generac. The PWRcell battery is a high-voltage system ranging from 360 to 420 V DC, and usable capacity ranges from 9 to 18 kWh in 3 kWh increments. We can dial in the 15 kWh with ease.

There are still a few more considerations before wrapping up this design.

1. A 15 kWh battery does not leave much room for future growth unless the original calculation included an extra 15% to 20% capacity.
2. Either PWRcell inverter satisfies the minimum 6,000 W requirement and allows for future growth.
3. The homeowner's budget will determine how to proceed with product selection.

"OK, but what if we needed triple the battery size for three days of backup with no sun? What would we do?"

Great question! We must look at stackability, and this goes for all considered systems. You can stack multiple PWRcell units to gain more capacity. However, the datasheet states you can only connect two PWRcell battery cabinets per inverter. This maximum size equates to 18 kWh × 2 = 36 kWh, so we are still 9 kWh short of our 45 kWh requirement. We have reached another decision point:

1. Reduce load (ha, right!).
2. Add another PWRcell inverter (with solar) and 9 kWh battery cabinet.
3. Manage expectations and recommend a backup generator (if the ESS supports it).

"But what if the homeowner already has solar? Can I still use this system?"

Unfortunately, no. You would have to swap out the existing PV inverter with the PWRcell. This is common for DC-coupled systems.

It is time to move on to grid-up applications!

TIME OF USE APPLICATION

As with the previous example, I will assume the sizing pieces are complete. These systems will charge the batteries when electricity prices are low and then discharge them when prices are highest, saving the homeowner money. Except for specialized applications, like VPPs or apartment-style storage, I think you are *wrong* if you sell a battery system to a homeowner without an accompanying solar system. I cannot stress this enough. This system will not have a way to charge the batteries when the grid is down, so you might as well say it does not have that capability. I'm sure many people (probably in sales) will tell me I'm wrong, but I stand by it.

"Whoa, wait a minute, pal. Didn't you say as long as you manage customer expectations, everything will be OK?"

Sure, but eventually, a dead battery acts as a paperweight, and homeowners frequently (and conveniently) forget your discussion about not running the hot tub during a grid outage. Let's look at a ToU application with two options. The only thing we need for this exercise is the customer's yearly kilowatt-hour consumption (found on their utility bill) and the peak sun hours for their location (located in peak-sun-hour charts).

OPTION A: NO SOLAR

Total annual consumption (kWh) = 13,800

Peak kWh = 1,200 per year (3.29 kWh per day)

Mid-peak kWh = 2,700 per year (7.4 kWh per day)

Peak sun hours = 4

We have a decision to make: do we size the battery for just the peak consumption, or do we include the mid-peak? Let's do both!

The peak-only calculation requires daily consumption to make things easier when looking at a storage system. Most utility companies provide a way to download your interval data. However, the CSV- or XML-formatted spreadsheet will not be easy to analyze. Give this data to a 13-year-old kid so they can create fancy charts. I will spare you the gory details of extrapolating a year's worth of data. However, a salesperson with an automatic software tool or someone with extensive spreadsheet experience can easily extract the relevant information.

The sum of the two peak periods is 10.69 kWh. The next, and last, step is to find a storage system to accommodate this consumption. If the homeowner has a central air conditioner or other large loads, you will need to find a compatible ESS that can handle the surge. You don't even need to look at surge requirements unless your utility also hits you with demand charges. However, shoring up these two values as we did with the previous examples results in a bulletproof system. Good news! After rounding our system requirement up to 11 kWh, a quick internet search reveals every ESS manufacturer has a solution that will work for this project or get really close.

See? Easy stuff.

OPTION B: WITH SOLAR

Sizing with solar for our 11 kWh consumption requirement, we would plug this value into the following formula. We will use the peak sun hours for ZIP code 01007.

Estimated PV-system size =
Total kWh per day / peak sun hours / 0.8[7]

11 / 4 / 0.8 = 3.43 kW

For this application, we will need, at a minimum, a quaint 3.43 kW PV system. It would be wise to size up since we want to charge the batteries, and the weather is not always accommodating. This sizing stuff is easy when you don't have to consider backup. Unfortunately, the harsh reality is you *do*! Every battery system will switch to backup mode when the grid goes down.

HIGH-VOLTAGE BATTERY SYSTEMS

This section will size a high-voltage battery system: SolarEdge's StorEdge using the LG Chem RESU10 battery.[8] I will only cover the self-consumption application. We will size all three modes the same way regardless of the AC or DC coupling architecture.

PROJECT DETAILS

Total watts—6,000

Total watt-hours—15,000

Total surge—9,600 W (40 A)

PV array—6,500 W

PV inverter—6,000 W

Average PV production—8,400 W

Location ZIP code—01007

The StorEdge unit acts as the charging mechanism and also accepts the PV array. However, SolarEdge does not make a battery, though they do pair well with the LG Chem RESU and the Tesla Powerwall. If there is an existing PV inverter, you must replace it with the StorEdge inverter. The StorEdge product is not compatible with a generator as of publication.

Since the LG Chem and Tesla products are so similar, I will cover both. The StorEdge SE7600A-US output easily meets the 6 kW power requirement. Hint: it's in the name of the inverter (7600). You can also three-stack the StorEdge. But it gets a little dicey when we look at the AC outputs of the battery systems. Although the SE7600A-US has a 7,600 W continuous output, the LG and Tesla batteries can only sustain 5,000 W.

StorEdge has two maximum AC-output ratings (surge): grid-up operation and backup. Unfortunately, at 8,850 W and 6,600 W, respectively, both ratings fall short of the required 9,600 W. We have reached a design consideration point for the AC-output requirements:

1. Stack units to meet the total power and surge requirements.
2. Manage customer expectations and tell them they will not be able to operate large loads.
3. Use a soft-start kit to reduce the surge to a reasonable level.

When considering battery size, the usable capacity of the RESU10 is 9.3 kWh, which falls short of the 15 kWh requirement. However, the battery is stackable (up to two batteries per StorEdge inverter), and adding one more will be more than enough. Although more expensive, the homeowner has a lot of room for growth. Of course, you can always inform the homeowner they will be a little short on capacity during those long, cloudy Belchertown winter months.

The Tesla battery has a usable capacity of 13.5 kWh, so the previous considerations apply. However, the battery can quintuple stack.

So there we have it. We have matched our sizing results to product selection. All that is left is the most important and most challenging part of the process—managing the customer's expectations of their new system.

CHAPTER 12

MANAGING EXPECTATIONS

Managing expectations is arguably the single most crucial thing you can do to prevent a poor customer experience with a solar-plus-storage system. I would wager that solar inverter and energy storage system manufacturers have a noticeable number of service calls related to awful homeowner experiences. And I am not talking about component failures or wear-outs but rather a homeowner's perceived "abnormal" behavior of their system caused by a lack of education or awareness of what their new system is supposed to do. A big problem in our industry is that some companies market to their customers, but they do not educate them. The lack of end-user training costs you money!

This is starting to sound salesy, isn't it? I would rather talk about FFT in high-end lighting circuits or IGBT switching frequencies or even perform a little theory-to-practice with some electrical circuits than talk about sales techniques. However, I will never forget what my boss, Dave Wojciechowski, VP of sales, SMA America, told me back in 2009 as I politely protested my shift from the service department to the sales umbrella as a technical trainer.

"Smitty, we're all in sales. As a trainer, you talk to more people than our salespeople could ever hope to reach. Embrace it."

My attitude quickly changed when I attended the sales team-building events. Driving Range Rovers on a private course in Malibu and sailing to Angel Island in San Francisco Bay for a barbecue almost made me consider a lateral move. It sure beat the nature walks and cooking classes I took when the technical training team shifted over to marketing. Lame.

My not-so-humble opinion is that solar-plus-storage salespeople need technical sales training to better sell ESSs. Improperly educated homeowners generally cause awful customer experiences in this (and any) industry. Most of the calls I am on, most of the service logs I read, and most of the anecdotal stories I hear at industry trade shows point to bad sales and marketing. Years of solar sales sheets have spoiled most of us, and it is a hard lesson to learn that selling storage is not so easy.

The sales process for solar has been rendered down to just a few talking points accompanied by a sales sheet. The sheet has three to four columns, depending on how ambitious the product comparison will be, and the column with the greatest number of green check marks wins! Interestingly enough, regardless of the manufacturer, the green check marks always seem to end up in their favor. Look familiar?

SALES SHEET FOR BRAND A

SPEC	BRAND A	PRODUCT X	PRODUCT Y	PRODUCT Z
OUTPUT	✓	✗	✗	✗
CAPACITY	✓	✗	✗	✗
SURGE	✓	✗	✗	✗
AC LOSSES	✓	✗	✗	✗
DC LOSSES	✓	✗	✗	✗
PV LOSSES	✓	✗	✗	✗
EFFICIENCY	✓	✗	✗	✗
GENERATOR	✓	✗	✗	✗
BLAH	✓	✗	✗	✗
BLAH	✓	✗	✗	✗
BLAH	✓	✗	✗	✗

These crafted sales sheets make Brand A look like the best option out there, and there is usually a statement accompanied by a few exclamation points reassuringly stating so. However, many of these sales sheets focus on various specs that don't really matter for grid-tied storage. It is also hard to compare and contrast storage systems since the specs are so different.

Selling solar has been boiled down to a reasonably successful formula depending on the area and how many incentives and rebates you can pile on. The initial doorbell-ring visit goes something like this:

knock knock / ring ring

Homeowner: "Hello?"

Salesperson: "Hi, I'm with <insert company>, and we have been installing a lot of solar in your neighborhood. How would you like to have cleaner, cheaper energy? I have the equipment, and you have the roof space. I'll pay for the installation, and then we can split the money we get from the utility. Sound good?"

It is this easy; however, internet-savvy consumers will research products and services on their own before making a purchasing decision.

"See? We have the most green check marks! What do I have to do to get you to take home one of these babies?"

—SALESPERSON

People have their own definitions of "best," and what is best for one person may not even register to someone else as an option. Frankly, I don't think there is such a thing as the "best" anything for this very reason.

"These microinverters are the *best* option for shaded roofs!"

"I don't have a shaded roof."

"But it's still the best option for your house!"

"Why?"

"Because of these other check marks!"

Or

"Hey, this storage system is perfect because it has the *best* round-trip efficiency."

"But the warranty states it only has 5,000 cycles. That one over there has 8,000."

"Yes, that is true. Let's get you set up with that one!"

Some salespeople use these sheets as a crutch so they won't have to gain any technical knowledge about the product. Don't you want to be able to address questions about a product you do not stock? If you don't, you are wrong. You should want to be the subject-matter expert because:

- Your credibility will increase.
- Your company's reputation will improve.
- You will appear to be a baller since you won't have to say, "Good question; let me get back to you on that" like so many other salespeople.

But we don't live in Should Land, do we? Storage sales are beginning to take on the same lackluster approach.

CONSEQUENCES OF A BAD CUSTOMER EXPERIENCE

Thirty-three percent of Americans say they will switch companies after just one instance of poor customer service. Most homeowners will not rip out their storage system because of poor customer service (although some have). Still, they are more likely to tell 15 people about their negative experience but only 11 people about their positive experience.[1] That could cost you future sales.

Consider the cost of a subpar service team when it costs 5 to 25 times more to acquire a new customer than retain an existing one. Take care of your customers or your competition will.

Two-thirds of buyers value the experience over the price. Give them a positive experience throughout the life cycle of the product, not just during the sales process!

Three-quarters of people think it takes too long to reach a customer-service agent. Here's what you can do to dispel that preconception:

- Answer the phone!
- Properly train your service team on all products.
- Incentivize your service team to stick around.
- Hire the right people to lead, manage, and mentor your service team.

Installers and homeowners can forgive products that prematurely fail, wear out, or experience an occasional hiccup as long as the service and support are there. This has been self-evident in the last 15 years as new solar and storage companies have come to market and jockeyed for position on the who's-who lists. Every solar and storage company will take its lumps with quality after a new product launch. The big difference, and equalizer, is how that company takes care of you. Service calls usually come in after the installers have been on-site for an entire day. They want to go home, something isn't working, and they are as frustrated as being on the receiving end of a double letter Q, double word score for "queen." Take care of them.

THE (STUPID) ROI ARGUMENT

A common objection to selling or offering storage systems sounds something like this: "The ROI for storage just isn't there."

No ROI compared to what? Some people collect fine art, fine

wines, antique cars, coins, or stuffed animals in hopes of a future profit or no profit at all. They buy this stuff for pleasure, personal satisfaction, or because it's just freaking cool. This objection seems lazy to me.

I was a hardcore PC gamer for a long time, and about every 18 months to two years, I would completely upgrade my system to keep up with the latest gaming requirements. I'll never forget my first home-built system back in 1993—an Intel 386 DX2 microprocessor with 4 MB of RAM, a 200 MB hard drive, and a 14.4 Kbps modem. Yes, that was cutting edge back then, and that system cost me about $2,000. I almost immediately upgraded my RAM with another 4 MB for $200. Today, I can get 32 GB of RAM for the same price, nearly an 800,000% increase in memory. I wasn't thinking about ROI. I just wanted to play *Mechwarrior 2* with a buddy of mine online.

Your customers have similar stories; we all do. People buy expensive things all the time without considering the ROI for one second:

- A $7,000 refrigerator with an opaque door that reveals the contents behind it with a quick two taps, a floor sensor that will auto-open the door after you swipe your foot under it, Wi-Fi capability that updates your shopping list after taking images of its contents, and all sorts of other must-haves
- $1,000 rear multimedia package upgrade to keep a minivan full of screaming kids entertained with a looping cartoon or video games for that long ride back home from the school pickup
- $150 to $200 for a shampoo and hairstyling
- $400 for quarterly Botox treatments
- $10,000 for a home entertainment system
- $20,000 for an F1 Savannah cat

- $1,000 for metronomical smartphone purchases
- Thousands of dollars per year devoted to new clothes, shoes, and accessories

ROI is different from *value*. Value is what you pitch.

Nobody seemed to care about the ROI of their solar system until the late 2000s, when the big-box solar companies came onto the scene and made solar available to the masses. Power purchase agreements (PPAs), leasing, and even renting solar systems brought solar mainstream. Sign on the dotted line, and the system would pay for itself in five to seven years. Then the rest is gravy. When those guys started shoving ROI in our faces as a sales tool, then we cared. It reminds me of the brilliant marketing campaign of Listerine. Until 1912, nobody cared about stank breath, but then Listerine convinced people it was a societal faux pas. *hat tip to* Freakonomics[2]*

DOLLAR PER KILOWATT: THE METRIC THAT COMMODITIZED SOLAR

When solar modules became a commodity, it wasn't long before the inverters followed. I cannot remember when I first heard it, but I will never forget when I finally reached my limit. It was at the annual Solar Power International show in 2012 (Orlando, Florida) when a man walked up to me at the SMA booth wearing a theme-park hat and carrying his swag bag and a handful of spec sheets. He asked me these two questions:

"SMA? You guys make inverters, right?"

"Yes, we do," I cheerfully said as I secretly nursed my hangover.

"What is your dollar per kilowatt?"

I looked at him, and with a smile, I pointed and politely said, "Oh, our competitors are thatta way."

When I saw the confused look on his face wondering what to write down on his notepad, I told him if dollar per kilowatt is all he cared about, he would not like what I had to say. The dollar-per-kilowatt conversation is as lazy as the ROI discussion is subjective. It is a double-edged sword trying to sell a system to fit the customer's budget instead of one they need. This premise is perfectly demonstrated by Mike Wight of Gardner-Energy in Utah. He has more experience than anyone I know sizing, designing, and installing off-grid systems. Mike forewarns that "A $2,000 to $5,000 budget will not get you a system that will work with the solar on your roof. I can't tell you how many phone calls I ended short just by asking what budget we had to work with."

COMMON SERVICE CALLS

How many times have you talked to an angry or confused home-owner about something they "should have known"? These calls are not to be confused with the presale misconceptions of:

- Solar doesn't work in cold/hot/cloudy/snowy weather.
- Solar and batteries are too expensive.
- Batteries are a pain to maintain.
- That stuff doesn't work.

Here are a few examples of poor expectation management and customer experience.

Call: "Why isn't my solar system making 6,000 W? You told me I was getting a 6,000 W system."

Cause: The salesperson did not explain the difference between nameplate value and reality.

Call: "I think I have a bad solar panel. This one is making 280 W, but the one next to it is only making 270 W." (You MLE installers know this one well.)

Cause: The salesperson did not explain the difference between nameplate value and reality.

Call: "You told me I was going to be able to ride out these utility-mandated power outages with this solar and storage system you installed, but I didn't even make it through the first day!" (They neglect to tell you they had a hot-tub party with their friends.)

Cause: The salesperson did not explain how large loads drain the battery bank.

Another cause: The homeowner was only half-listening to the salesperson and didn't connect large loads with battery drain.

Call: "Why didn't you tell me my solar would not work when the grid goes down?" (Yikes! This call is all too familiar.)

Cause: The salesperson did not explain that solar does not work when the grid goes down and a battery system is the only thing that will make it work.

Call: "Why does my solar monitoring say it is making 4,500 W, but the battery-system monitoring says 4,200 W? Something isn't right. Come fix it."

Cause: The salesperson did not explain the variances in the manufacturer's measuring tolerances.

Another cause: Installer incorrectly installed measuring components.

Call: "You said I was getting a 10 kW system, but I see the spec sheet says it can only put out 5 kW."

Cause: The salesperson confused kilowatts with kilowatt-hours.

Another cause: The homeowner confused kilowatts with kilowatt-hours.

One of the most memorable customer-service calls was a lady calling about her PV inverter. This particular model had built-in Bluetooth for monitoring. This woman made her rounds talking to different people in service, sales, and then whomever she could reach. Her complaint was one of the most fascinating inferences of power transmission we had ever heard. She had convinced herself her neighbor was stealing her solar power via Bluetooth. She stated the output of her inverter never reached the 2,500 W printed on its label. Everyone explained how modules work and the influences the environment and the weather have on production. She wasn't buying it. It seemed more plausible to her that her neighbor could somehow steal energy via radio transmission.

I recall another service call I took early one morning concerning a dead storage system. This off-grid customer had purchased a solar and storage system but did not have a generator on-site. They drained the batteries down to the point the system shut off to protect against a deep discharge. Sadly, their 3 kW solar system was not sufficient to charge the batteries in January at their location. They left their home, boarded their pets, and checked into a hotel. They were there for over a week.

Although this mishap was an off-grid application, this could very well happen to someone with an extended grid outage. The utility company has at least three or four backup options to supply power to their customers. Why wouldn't you put that same level of redundancy into an ESS?

I could write volumes filled with anecdotal references, but perhaps that will be another book. The underlying message is always the same—you have to talk to the customer and explain realistic system behavior. And it helps to install a system that is appropriate for their application.

"Sure, but some solar professionals don't know what they don't know. How are we supposed to figure this out?"

Let's leave that to the next section.

CHAPTER 13

STREET CRED

There are a few things that can significantly increase a homeowner's satisfaction with their solar and storage system. These three rules of thumb can best be summed up by the famous quote from *Fast Times at Ridgemont High* when Brad Hamilton told Jeff Spicoli to "Learn it, know it, live it." Anyone good at anything has conquered these three commands.

1. LEARN IT

While it is not reasonable to expect someone outside the subject-matter expert's title to know every detail, there are some basics everyone should know. These industry basics are familiar to most people who are around the products every day: the installers who put it all together and then go back out to the job site when something goes wrong and the designers who figure out an accommodating system size for the homeowner's load demand and then juggle those demands with reality. And then we get to sales.

I have seen some good salespeople, some bad ones, and some who couldn't find their ass with a flashlight. The good ones will help the customer make an informed decision. They may not know the ins and outs technically, but they can navigate the spec sheets and understand system operation as a whole. A lousy salesperson uses the spec sheets as a crutch. They read from them with a slight nervousness in their voice. They say, "I'll have to get back to you on that one" too many times during the conversation, or worse, they just make stuff up. They don't know how it works, how it integrates, or anything off their sales script. They need sales sheets with green check marks and red Xs for product comparison since they do not have the necessary knowledge to make that decision objectively. Let's trim this 16-pound brisket one cut at a time.

SALES AND MARKETING SHEETS—WHAT MATTERS AND WHAT DOESN'T

Product sales and marketing sheets do contain some technical information for designers and salespeople. News flash! It is usually the info that makes their product look good. Product manufacturers take those specs and create handy comparison charts for salespeople to push. Evaluate specific specs between manufacturers for any piece of the solar and storage system. Although I will not present a spec sheet, you can find them online within milliseconds using your favorite search engine. Let's decipher them by groups, starting with solar inverters.

Solar Inverters—Direct Current Values

Maximum PV power—The recommended largest size of the solar array.

It is common for designers to size the PV array to a larger size than the inverter. In a previous chapter, you learned inverters spend about 80% of their lifetime producing 50% of their nameplate rating. To compensate for this, residential solar designers will use a 1.1:1 to 1.5:1 DC to AC ratio. A 5 kW inverter will typically have a solar array sized between 5,500 and 7,500 W.

This oversizing is not an issue, and the inverters can take it. In fact, you cannot hurt a PV inverter with too much power. If you could swing the wire management, you could connect a 20-kW array to a 5-kW inverter. The inverter will only produce 5,000 W, but it will get there pretty fast in the morning! This inverter would not show a gentle bell curve but rather an abrupt square wave on the monitoring portal or app. This oversize is a tremendous waste of modules (in both cost and roof space) since the PV inverter could never hope to use all that DC. Plus, an inverter running at full power day after day, year after year would probably have to be serviced more often. Power electronics can only take so much. Imagine running your car engine at maximum horsepower every time you took it out for a drive.

"Hey, so what will damage a solar inverter, since a huge array won't do it? Why does this spec exist, then?"

Great question! The spec is a courtesy—a gift from the manufacturer that balances the PV's cost with proper inverter operation. Something else the PV inverter manufacturers typically provide is a string-sizing tool. It dictates the number of strings and how many modules per string make an acceptable array size constrained within the inverter's operational specs—mainly the 600 V limitation dictated by the NEC. The string-sizing tool will display the maximum voltage somewhere in the 580 or 590 V range. Adding one more module would bump the string voltage over 600 V, albeit perhaps not by much. Designers would call with this question, and I had to give them the official answer—don't do it.

High DC voltage targets the inverter capacitors. They look like a soda can left in the freezer for too long. Some explode. Some will just show deformation. Regardless, warranties do not cover this type of damage. If the inverter is sent back to the manufacturer for service, the leaking electrolyte is a tattletale for this violation. There is also a tattletale in the firmware that records the highest voltage ever measured by the inverter and how many times the array exceeded the upper threshold.

"Aw, come on, man. The inverter would blow up the first time that voltage went over 600?"

No, those capacitors have a higher voltage rating, but continually stressing them will eventually cause damage. Don't do it. There is always an outlier story, though. I recall such a story involving one of the first SMA America inverters on the US market.[1] One customer had it installed on his home nestled in a valley with snow-covered hills on each side. On a cold, crisp spring morning, the sun finally rose over those hills at about 10:00 a.m. The modules were still cold. When that full sun hit the array, the irradiance skyrocketed because the snow-covered hills acted as a reflector. It was the perfect storm. The voltage exceeded the rating of the capacitors and then blew up.

One of the repair guys told me that electrolyte was leaking out of the unit as it sat on the workbench. He connected the tattletale cable to the inverter and confirmed what he already knew. The repair person found out that SMA's service line did help this customer with the sizing. The sizing tool correctly accounted for all environmental variables except those snow-covered hills and the one cloudless day during that time of year. How could it, right? SMA did the right thing and sent a new unit at no cost, but there were some serious discussions about the array size.

Maximum power point tracking (MPPT) window—This value

represents the manufacturer's recommended voltage range that will allow the inverter to operate at its most efficient power range.

This spec may have various names, but the telltale value will be a voltage range. A range such as 150 to 480 V, for example. It is also a value deserving of some bragging rights. Wider windows are usually indicative of better engineering. However, some manufacturers may widen the window for marketing reasons. The low end of the window translates to more production at lower irradiance levels. I would argue if the window is too wide, these low light levels aren't providing any real advantage. All it takes is a small fluctuation in irradiance (moving cloud cover) to cause the voltage to fall below the window lower limit and the inverter to shut off. Are those few watt-hours of energy worth the future replacement costs because the inverter works a bit longer in the day or turns on 15 minutes before another manufacturer's product? Probably not.

The MPPT window may have two values—a rated and an operating range, the latter having the wider window. Not to worry, most string-sizing tools will automatically keep the voltage within the rated range.

PV start—The voltage at which the PV inverter will attempt to start making power.

Usually applicable in the morning as the sun begins to rise, this value is inside the MPPT window. It should prevent accidental shutoffs, previously discussed, but only for a well-designed or engineered system. Commonly used as a sales and marketing perceived advantage, it really isn't. I have yet to meet an installer who based their purchasing decision on the MPPT window or PV start spec. It's akin to establishing a car purchase on which model idles the slowest.

Short-circuit current—The maximum amount of current the unit can accept from the solar array. Solar inverters are current-limiting devices (see the section on MPPT). As previously stated, you cannot damage an inverter with too much power, only voltage. Inverters automatically throttle down the solar array to an acceptable level of current. However, overcurrent protection, like fuses and disconnects, rely on Isc when things go wrong.

Solar Inverters—Alternating Current Values

Power—The maximum amount of watts the unit can produce.

Typically, there are two types of listed power: nominal and apparent (or maximum). In most grid-tied residential inverters, the values will be identical. If not, go with the nominal value. The relationship between the engineering aspects of real, apparent, and reactive power is not essential to the homeowner or even most industry people. However, they become critical talking points in larger commercial or utility-scale solar systems.

Larger nameplate values equal more harvest. An 8 kW inverter would typically produce more energy than a 5 kW inverter. Using PVWatts,[2] a 5 kW system (using only default values in the tool) will harvest roughly 7,800 kWh per year in Sacramento, California. This same system in Denver, Colorado, would crank out 8,000 kWh per year but only 5,800 kWh in Syracuse, New York. The average annual consumption for an American family is about 10,000 kWh per year. None of these PV systems would completely cover this consumption, but it will get them close.

Sidebar: PVWatts is a free tool sponsored by the National Renewable Energy Laboratory (NREL) and the Department of Energy. If you are ever near Golden, Colorado, and have the opportunity to visit NREL for a tour, do not pass it up! I went with a few SMA colleagues back in 2009, and it was incredibly informative. We saw the PV-module testing area full of different PV arrays, some of which had been there for 30 years. The tour guide explained that some of the arrays suffered less degradation than the industry-accepted rule of thumb of 1% per year. It makes a compelling case for module quality.

Voltage—This spec represents the various grid voltages compatible with the model.

Almost all residential is 240 V, with a few 208 V homes scattered around. Nearly all PV inverters are compatible with both voltages. The inverter will automatically detect the voltage. There is also a range associated with these voltages that will allow for voltage drop or rise. The tolerance is typically –10% to +12% of the nominal grid voltage. If the inverter detects a high or low out-of-spec voltage, it will disconnect from the grid. Suppose the voltage falls outside this window. In that case, the inverter will automatically stop making power but attempt to reconnect at five-minute intervals.

Frequency—The grid frequency compatible with the inverter.

Some US models will support not only the native 60 Hz but also 50 Hz as well. This dual-frequency feature allows for greater flexibility in Mexico, the Caribbean, and Latin America, where the grid frequency is 50 Hz. There is an operating range to account for a grid that may fluctuate. Typically, the tolerance is 59.3 Hz to 60.5 Hz for North America. If the frequency falls outside this window, the inverter will automatically stop making power but keep checking at five-minute intervals before reconnecting. Sometimes the frequency window is widened, as approved by the utility, for areas where the grid frequency may meander in and out of this tolerance. Puerto Rico and Hawaii are two examples of this.

Output current—The maximum amount of current the inverter can produce.

This is another engineering spec for overcurrent-protection calculations. While it is tempting to divide the output power by the nominal grid voltage to find the output current, this might not always give you the inverter's actual value. You must use the spec sheet to find the manufacturer's values. For example, a string inverter lists the output at 6,000 W, and 6,000 / 240 = 25 A. However, the spec sheet states the output current is 25 A when connected to a 208 V service.

"Huh? But how can the output current be the same if the voltages are different? Six thousand divided by 208 equals 28.8 A."

Correct! But in plain sight, the manufacturer's spec states the maximum output power for this inverter on a 208 V service is not 6,000 W; it is 5,200 W. 5,200 / 208 = 25 A.

"Oh, so they are ripping me off of 800 W? Great!"

There must be compromises when it comes to power output and overcurrent protection. A 25 A value means the installer must use a 35 A breaker (max current times 1.25). However, sometimes these in-between sizes are not readily available. The installer would then use the next higher size of 40 A. A PV maximum inverter output current of 28.8 A means using a 40 A breaker. Breaker sizing is the reason manufacturers make oddball inverter sizes with outputs like 7,600 or 11,400 W. The 40 and 60 A breaker sizes, respectively, are much easier to find. Also, microinverter branch circuits are usually limited to 16 A total, regardless of how many micros. Most microinverters will require a 20 A two-pole breaker per branch circuit. Each branch will have a nominal output of around 3,840 W.

Power factor—A ratio of the working power to the actual power required to run a load.

It is a measure of how well the PV inverter can produce power relative to the grid power (in simplest terms). The Greek letter Phi—$\cos(\varphi)$—is the symbol for power factor. A power factor of 1 indicates the inverter will sync to the grid voltage and frequency at parity. Most PV inverters will have a power factor of 1. However, some will also have a range, usually 0.8 leading to 0.8 lagging. They will produce power to compensate for inductive or resistive loads accordingly. Again, only engineers care about this spec, and most normal people can mostly ignore it for residential applications.

Total harmonic distortion—Another engineering spec that provides a measure of the power quality.

Mostly ignored by most people (because almost all inverters have the same value), but required by certification authorities. Anything less than 5% total harmonic distortion is acceptable.

Efficiency—A measure of the DC to AC conversion.

As inverters convert DC to AC, there are some efficiency losses due to heat transfer. Very few devices are 100% efficient in this universe. Our sun is only 0.7% efficient at converting hydrogen to helium! Most motors are about 80% efficient at 75% load. Car engines are about 20% to 35% efficient. Commercially available solar cells are topping out around 14% to 22% these days, although the Mars rovers' cells are 29.5%. Solar inverters are pretty efficient compared to these devices, with even a bad one hitting 96%. String inverters are around 98% for a transformerless model.

Manufacturers will list this value as "peak" efficiency, which means a laboratory-tested value. Out in the wild, these inverters may rarely ever hit that peak. The California Energy Commission (CEC) provides a weighted average that considers six different systems and environmental efficiencies. It creates a level playing field for all manufacturers trying to get a marketing edge with their efficiency value.

Another thing to note is the inverter efficiency will vary according to its output power. PV inverters are typically less efficient at lower and higher output levels. Almost all manufacturers will provide an efficiency chart to show this. However, some are better than others. An efficiency curve that is relatively flat after a moderate power level shows some serious engineering.

The rest of the PV inverter spec sheet includes the weight, dimensions, outdoor rating, and so on. The outdoor rating isn't too relevant since most PV inverters have outdoor ratings. I have never seen one that doesn't, but I am covering my *okole* just in case.[3] You might hear rumblings about the NEMA rating, which designates a protection rating against the intrusion of liquids, rain, ice, corrosion, and other contaminants.

Almost all PV inverters have, at a minimum, a NEMA 3R rating, which means the device can withstand typical weather conditions. However, these devices are not watertight and should avoid constant sprinkler hits.

NEMA 4X-rated inverters and devices have corrosion-proof materials and can withstand falling dirt, rain, sleet, snow, windblown dust, splashing, and direct sprayed water.

NEMA 6-rated devices offer the highest protection and provide a degree of protection against water entry during occasional, temporary submersion at a limited depth. Almost all microinverters have this rating.

I recall a 2009 solar event in a crowded auditorium when a panel discussion derailed into a microinverter lovefest. Someone stood up and emphatically stated, "String inverters could never survive microinverter testing!" The vibe was comparable to a Beatlemania frenzy. This statement and crowd reaction still puzzles me. It is like someone standing up during an automobile panel discussion and, while shaking their fist at the sky, yelling, "A Ford Focus would *never* be able to run the Baja 500!"

Of course a string inverter would never survive those tests. One can take comfort in the fact that they don't have to! Microinverters live in the system's most inhospitable location—underneath a sheet of glass that bakes in the sun. It stands to reason they should have different testing standards. However, most US string-inverter manufacturers recommend a mounting location that is not in full sun for most of the day, that is, the house's south side. This location will cause the inverter to run hotter and, therefore, less efficiently. It could lead to premature servicing or, at the very least, a replacement of the LCD.

One more thing: unless you like waking up early in the morning to a buzzing noise, I would not install a PV inverter on the outside of a bedroom wall. When the sun comes up and the inverter is producing at a low power level, a few hundred watts or so, the inverter will buzz. Although normal behavior, it turns into an annoyance when trying to sleep in on the weekends.

Battery Inverters

Battery inverters share similar specs with PV inverters, so I won't spend too much time on the duplicates.

Battery Inverters—Direct Current Values

Input voltage—The rated battery voltage the system needs to operate.

Make sure your batteries match the battery inverter requirement.

Voltage range—This spec indicates the lowest voltage to keep the battery inverter operational and the highest voltage that, if exceeded, could damage.

Battery voltage is not constant and will fluctuate with charging and discharging. This spec will represent the range based on the input voltage. For example: a 48 V inverter = 41 to 63 V. Design tools will help keep these voltages in safe zones.

Max charging current—Highest amount of current the inverter will use to charge the batteries.

Higher current means faster charging times, but the lithium-ion battery-management systems manage this.

Battery Inverters—Alternating Current Values

Apart from the PV inverter nominal and maximum AC specs discussed earlier, storage products will have different specs that demand attention. Most show a higher-than-nominal output rating for a smidge more output power or provide surge power for larger loads. For example, the spec sheet may look like this:

Nominal power: 8,000 W

30-minute rating: 9,000 W

5-minute rating: 12,000 W

3-second rating: 15,000 W

1-second rating: 17,000 W

Efficiency—Same discussion as before but with one caveat. Battery inverters will also show a roundtrip efficiency value. Since battery inverters must convert DC to AC and vice versa, there are two conversion losses. So a battery inverter that is 95% efficient will have a roundtrip efficiency of:

95% × 95% = 85.7% (roughly)

You lost some energy when it went into the battery, and then you lost a little more as it was taken from the battery and converted to AC to power your home. Some manufacturers boast roundtrip efficiencies in the high 90s. Some salespeople push this spec as the only consideration since it will have a green check mark. I submit that this value doesn't matter in the long run for a few reasons.

First, it depends on how the system uses solar and when it supplies power to the loads. If you have an AC-coupled architecture, the loads will directly consume that PV production with only a 2% loss (for a transformerless inverter). That is on par with DC-coupled systems that boast these higher roundtrip efficiencies, usually greater than 98%.

Second, suppose you are powering most of your loads during the day, which most people will. In that case, the AC coupling will pretty much equal the DC coupling advantages.

"OK, but what if there isn't enough sun to power those loads? AC coupling doesn't really do much then, does it?"

No, but neither will a DC-coupled system. The array still needs sun regardless. This doesn't mean there are no useful applications for DC coupling. If most of the loads are at night, DC coupling is the way to go, primarily if you use a renewable-energy source like wind or micro-hydro that doesn't rely on the sun. Most off-grid installations use some form of DC coupling since these systems are experts at curtailing excess PV.

Lastly, when comparing roundtrip efficiencies, it is helpful to look at other specs, like the number of cycles. I have had more than one conversation with vendors and installers about this.

He: "This thing has 10 points of roundtrip efficiency on your product!"

Me: "This product does have a lower roundtrip efficiency. However, it has a higher cycle count. What good is a higher efficiency if you just have to replace the batteries more often? Any financial advantage of a 98% or 99% efficiency ends up flushed down the toilet."

Of course, none of this matters if you need a specific roundtrip-efficiency value for a rebate or incentive program. Even so, as long as it meets the minimum standards, the system is good to go.

COMMUNICATION

No, this is not a sales lesson on how to communicate with customers but rather a section on equipment communication, specifically monitoring the system using manufacturer or third-party devices. Every solar and battery manufacturer offers system monitoring, and Americans expect this type of visibility into their investment.

I have seen people use an inferior product in this age of app-based everything because of its kick-ass, intuitive, and remote-control functionality. I have also seen the opposite, and I completely understand. Who wants to use a poopy app with limited functionality that doesn't adequately display the information of their kick-ass storage system?

These systems are more than just cold, unfeeling equipment. ESSs become as important as an air conditioner in Tucson, Arizona, or as critical as a backup generator in an Arkansas backwoods fishing cabin. They provide peace of mind, energy security, and a way to keep medicine refrigerated and food (beer) safe. These renewable-energy systems have become a way of life for some. To others, perhaps they are just something that takes up space in their garage, but none of them dismiss the importance of monitoring their system or showing it off to friends and family.

"OK, we get it. It's important. Get to the point already!"

Rude! Can you guess the number-one call to any manufacturer's service line? When I ask this question during training, the resounding answer is usually the correct one. It isn't a hardware malfunction or a software issue (although software is a close second). It isn't operator or installer error (although those are pretty high on the list). It is communication—specifically, a lack thereof. The system is not communicating with the monitoring devices.

"Um, yeah. I'm in my summer home in the Hamptons, and I cannot see what my system is doing back in California." (Actual call from a homeowner.)

The main reason solar and storage systems do not show monitoring data is a disconnection from the internet. I would bet the people who have the most issues are not using a hardwired connection from the router to the battery or PV system. There are many reasons: the house suffered a momentary power loss, someone tripped over the router power cord, the ISP sustained a loss of service, and so on. The reasons are endless and would affect any device regardless of its connection—Wi-Fi, ethernet hardwire, PLC, and so on. But the first thing the service line will ask is *how* their unit connects to the customer's router. If you tell them anything other than a hard line, you might be in for a long day. Or maybe not. Perhaps a system restart will take care of it, and all it cost you was a truck roll! Considering the average truck roll costs $400, it is an expensive trip to reboot a router. Generally speaking, if the service tech cannot remotely log into a PV or storage system, there is something in the customer's network preventing a solid and spicy connection. Check the ethernet cable, firewalls, ethernet adaptors, router settings, and so on.

In all fairness, I am not taking my own advice. My ESS is in my garage, and my router is at the opposite corner of my house, which does not have a crawl space. There is no way my wife would let me run blue ethernet cable across the living room, the dining room, and the kitchen and through the wall into the garage. I bought a TP-Link AV600 PLC adaptor, and it works fine. I plugged the base into the router and nearby outlet. The extender is plugged into the same receptacle as my irrigation controller about five feet away from my ESS. When my power goes out, it will take five minutes for the PLC to reconnect, even though my ESS flips over pretty fast. My router is now on a small UPS that also keeps my desktop computer on during the switchover from the grid to the microgrid. Some homeowners will not be as hands-on, though.

GENERATORS

Common question: "Do I need a generator with this ESS?"

Answer: It depends!

- Is there room for one?
- Will the homeowners' association allow it?
- Is the ESS compatible with generators? Some are, and some aren't.
- Are there extended power outages with frequent consecutive days of bad weather that would inhibit battery charging?
- Are there large loads that will necessitate higher-than-usual battery cycling?
- How will the homeowner handle an unexpected outage when the PV may not be producing enough to charge the batteries?

The last one, of course, is rhetorical. The owner will freak out because they paid tens of thousands of dollars for a system that didn't keep their lights on. I am a big fan of generators and will always recommend one. It could be a small-output model, just enough to get the batteries charged during a rough outage. Areas prone to severe weather should spec in a generator (and fuel reserve). Period.

Common sales question: "Why would I want to install a solar and storage system and make $3,000 when I can install a whole-home generator, which is easier and doesn't take as long, and make $6,000?"

Answer: Maybe you wouldn't. But this also depends on what the customer already has. If they do have solar, it is useless when the grid goes down. Do not install a PV inverter on the load side of a generator unless you like replacing regulator boards. So, the pitch would be something like, "You have all that solar up there. Wouldn't you like to optimize that system by adding a battery and using stored solar at night?"

GRID DEFECTION

There is a lot of talk about grid defection and how the projected increase in utility prices can eventually make it an economic reality. You will, inevitably, come across one of these customers. The Rocky Mountain Institute's 2014 report *The Economics of Grid Defection*[4] lays out the financials very well. For example, the residential solar and storage grid-parity chart estimates that by around 2037 in California, the cost of solar and batteries will decline to the same cost as utility prices. Sticking it to the man is a motivator for many people. Ask anyone in PG&E territory. It is a sexy and romantic idea to tell the utility to come out and remove its meter, but it is also a pipe dream for the overwhelming majority of utility customers. And here's the simple reason why:

GREG'S AXIOM #5

Most people cannot adjust their lifestyles to live off-grid.

Along with roof-space requirements, this axiom will almost always prohibit grid defection from reaching the mainstream, at least for most residential applications.

Let's look at a hypothetical. Comparing yearly consumption to yearly PV production shows a deficit of 4,000 kWh, meaning there is not enough sun to charge the batteries. We must spec in a generator since there is no grid to supplement this deficit. A 10 kW generator uses about 100 cubic feet of natural gas an hour at half load (5 kW). A SWAG (scientific wild-ass guess) calculation suggests 800 hours of run time a year to cover 4,000 kWh at 5 kW of consumption per hour.

Natural gas is a cheap and common fuel source. The Rocky Mountain Institute report does an excellent job selling residential defection even with these considerations in areas where utility prices are highest. The report focuses on Hawaii and the Southwest for many of the examples, charts, and graphs. Hawaii is a good starting point since most people do not treat air conditioning, if they even have it, as a life-or-death necessity, as they do in the desert Southwest. However, add an electric vehicle to the scenario and the ESS could have difficulties charging.

But it isn't an impossibility. I can look past my curmudgeonly attitude for a bit and agree that commercial applications will be the place to start. Solar and battery pricing is much lower than residential, and that is where investors are looking anyway. Most residential customers could realize a significant drop in their utility bills if they:

1. Change their consumption habits
2. Swap out all lighting and appliances to higher-efficiency models

Considering the average stay time in a home is 10 years, will owners spend the money to realize an ROI they might never see? Like most ideas, grid defection for residential areas will start with the early adopters.

2. KNOW IT

It takes more than *learning* everything in the previous section. You have to *know* it. This type of intimate knowledge can only come from extracurricular study time. But it will pay off in the long run because this knowledge will:

- Establish your credibility
- Allow you to speak honestly about the product

I will tell you from experience that the homeowners will appreciate it. I'm not advocating an abrasive, casually profane, tell-it-like-it-is-at-all-costs attitude (like mine), but forthrightness will go a long way. Knowing how the equipment works and interconnects with other system components will give you the confidence to speak to people using facts instead of making stuff up because you didn't know the answer. Attend vendor training as much as possible—and put your phone and laptop away! The trainer is the expert, so take advantage of his or her time. However, what you don't use, you lose. You must keep up with the products and any updates that roll out. Enough said.

3. LIVE IT

I have received many phone calls and replied to numerous emails and texts from people who are frustrated with system operation or performance. Their battery system shut down "for some unknown reason." The batteries are not getting a full charge from solar before the sun goes down, and then the power turns off in the middle of the night.

"How are we supposed to know what the homeowners expect the system to do?"

Ask them! If you look at their load sheet, assuming you use a detailed one and not an oversimplified online questionnaire, you will get an idea of what they want to do. For example, I will never forget a load sheet with a 100 kWh (2,083 Ah) per day requirement. I thought I saw an extra zero. Scanning the list, I noticed "construction equipment" accounted for 36 kWh and ran for 24 hours every day.

"Oh, yeah, um, that's actually my grow lights. It's my hobby."

One more detail: this person wanted to go off-grid. He lives in Oregon, and his solar requirement alone was 75 kW. His generator would need to run at least 24 hours per year. I directed him to traditional battery inverter / charge controller setups using lead-acid batteries. He would need about 4,166 Ah per day of capacity since we only discharge lead-acid to 50% DoD. Rough sizing his system required 72 batteries. I told him he was looking at around $27,000 just for batteries. And then he would replace them in about 7 to10 years.

"Does that include three days of autonomy?" he asked.

LOLOLOLOLOLOLOLOL (I thought to myself)! Oh no, my friend. You just tripled your battery cost! While an extreme anecdote, it is a familiar story. Here are some better examples for grid-tied applications. I have been on my fair share of three-way sales calls with the installer and the homeowner that go something like this:

Installer: "I see you have two pool pumps and an air conditioner on this load list. For self-consumption or ToU modes, this is not an issue. While we may not eliminate your utility bill, we will offset it by XX%.

"However, this system does provide backup capability if the grid goes down. These large loads will significantly reduce your solar and storage system's ability to deliver long-term power."

Into the rabbit hole we go!

Homeowner: "But the website said I only needed four of these batteries. Why wouldn't that work?"

Sales: "Because that online calculator was for a grid-up application and the tons of footnotes probably mean a backup application is not a good idea."

Homeowner: "Can't we just use a generator, then?"

Sales: "Unfortunately, this model does not support a generator."

Homeowner: "Can't we just install more solar?"

Sales: "Your roof can only handle 8,500 W of PV. The calculations say you need at least 20 kW."

Homeowner: "Well, then I want to go with Brand X!"

Sales: "That brand has the same limitations. Almost all of them will."

Homeowner: "Then find me one that doesn't."

Sales: "OK, this one can fulfill all your expectations. You will need a ground-mounted solar array. This is how much it will cost."

Homeowner: "What? That is way outside my budget. What can I get for my budget?"

Sales: "I just told you."

//fin//

I often turn people away because I know my product is not a good fit for their application. Salespeople need to learn how to do this. Ask yourself this question: is the sale worth the cost and hassle of repeated service calls or the "Hey, how come this is doing this?" customer call once or twice a week because of a working-as-intended system?

A bad customer experience is not worth it.

GREG'S AXIOM #6

No salesperson should sell a system and no manufacturer should offer a product less than 15 kWh.

It will result, more often than not, in a poor customer experience. For example, here is another all-too-familiar conversation.

Homeowner: "I really like that smaller-sized system. It fits in my budget."

Sales: "It does. However, it will not give you much backup time. Here is an infographic of everyday household items and how much energy they consume in one day. You said you wanted to back up your refrigerator, which typically takes about 1.5 to 2 kWh per day. That doesn't leave much for anything else."

Homeowner: "That's OK. I only wanted the fridge and some lighting anyway."

Sales: "What about devices like cell phones, laptops, and tablets?"

Homeowner: "Oh, yeah. Well, those don't take a whole lot of power to charge, do they?"

Sales: "No, they don't take a lot, only about 175 kWh per year each,[5] but you don't have a lot of capacity left over for other devices."

Homeowner: "Oh, what about the microwave? That can't be much. Oh, and I do have a mini wine fridge. And I like to play my electric guitar. Oh, and I do work from home, so I would need my desk backed up with the computer, monitor, printer, and all that. Plus, my wife and I were talking about getting an electric vehicle…" and on and on.

GENERATORS

Generators cause a lot of service calls. I know plenty of solar installers who shy away from them. Conversely, I see many of them go full bore and try to include them with every sale. These companies sold generators before they got into solar, so they know the intricacies. The common pain point is generator compatibility. Older generators can cause problems since their power may not be clean enough for the ESS to sync.

Things to look out for:

1. Ensure the generator output voltage and frequency match the ESS requirements. Some ESSs have tight windows, and if the generator falls outside this window, the ESS will not connect. For example, suppose the ESS requires a frequency between 59.3 and 60.5 Hz. In that case, an older generator that drops

below 59.3 Hz will not be compatible. The datasheets should provide these values.

2. Check the generator starting method. ESSs typically support dry-contact 12 V or 24 V signals.

3. Check the generator-starting mode. Most whole-home generators are auto-starting, meaning once the grid goes down, a signal tells the generator to start. The automatic transfer switch (ATS) usually sends this starting signal. If there is an existing generator, you will need to disable the ATS's auto-start feature. The default auto-start feature defeats the purpose of spending all the money on an ESS unless the homeowner doesn't mind the generator as an audible alarm for a grid outage.

4. If the existing generator is a manual start, the homeowner must understand and follow the battery charging procedure. It will be more involved and user intensive than using an auto-start generator.

5. Never put solar on the load side of a generator. Excess PV can destroy the generator board, and no warranty will cover this damage. Some ESSs have a way to isolate the generator from the PV, but some don't.

6. If possible, always use an auto-start generator with a compatible ESS. It will make things so much easier for the installer and the homeowner.

7. The homeowner must understand fuel-source accessibility. Generators are a lifesaver, but there might not be anywhere to refuel when you run out of gasoline or diesel. Filling stations may not have the power to run the pumps during an extended outage. Propane and natural gas are not as problematic. However, suppose you live in an earthquake-prone area. In that case, the first thing the utility company will do is shut off the gas as quickly as possible during or after an earthquake.

SOLAR INVERTERS

Although solar has become mainstream in many parts of the United States, these points bear repeating:

1. Solar does not work when the grid goes down. Otherwise, excess solar could backfeed into the grid and kill a lineman working to restore power.
2. Solar inverters rarely produce their nameplate AC rating. Most spend 80% of their lifetime generating 50% power. Explaining this to the homeowner will prevent "Hey, why is this 5,000 W solar system only producing 1,500 W?"
3. Module-level monitoring is a blessing and a curse. Explain to the homeowner that modules will not produce their nameplate value unless conditions support it. This explanation should prevent "Hey, why does this module say 160 W but the one next to it says 165 W? I have a bad module. Please replace it."
4. Explain to the homeowner that solar will produce a lot in the summer and not so much in the winter. Depending on system size and the customer's utility bill, a new solar installation may not cover the seasonal use. I have heard many conversations about homeowners convinced their solar system was not working correctly. They did not account for their increased consumption in the summer (air conditioning) or the low solar irradiance in the winter. The system was working as intended!

These are the most common concerns, comments, and system criticisms, with variants scattered about. Be ready. Fight the good fight!

CHAPTER 14

THE FOUR HOWS

Most people know about the five Ws of crime-solving (Who, What, Where, When, and Why). The four Hows should be a priority when managing homeowners' expectations. Here are the four basic questions customers ask (or should ask) during the presale conversation about their new ESS. Be ready for them:

1. How much does it cost?
2. How long will it last?
3. How long can I power my house?
4. How does it work?

Many homeowners don't know what they don't know. It is your responsibility to educate them, unless you just like losing money on unnecessary service calls. These four questions may seem obvious. But if they are so elementary, why do we keep talking to end users after their first power outage? Obvious answers are not evident to everyone because they aren't thinking about the right questions.

1. HOW MUCH DOES IT COST?

It is no coincidence that I begin with this question, so let's start peeling back those layers and talk about what the customer wants to know.

Sticker price should include parts, pieces, and labor and a statement that will allow for additional unforeseen installation costs.

Unforeseen costs could include rewiring of existing components not related to the current install. Many times, a committed do-it-yourself homeowner is the cause of rework. If you find something that isn't safe or needs rewiring, it will be on your dime if there is no stipulation.

WARRANTY

Everyone should know the manufacturer's warranty and coverage details. Here is the short list:

Product coverage—Know the warranties for the manufacturer's products since they could be different even for the same products. For example, newer models may have different warranty conditions than older ones. A PV inverter may have a 10-year warranty, but the battery inverter will only have a five-year warranty, even though both products are from the same manufacturer.

Performance—Including, but not limited to, the number of cycles to a specific DoD and the associated megawatt-hour throughput. You should be able to answer these questions:

- What does the manufacturer consider a full cycle?
- What is the DoD? SoC?
- Why is throughput important?

Manufacturer and installation-company obligations— It is essential to know who is on the hook for the system's components.

- Who pays for service visits?
- Is there a service reimbursement?
- Are there any homeowner responsibilities?
- Who is responsible for equipment damage during transit?
- Is a warranty extension available?
- What are the limitations?
- Are lightning strikes or other forms of force majeure covered?
- Are utility spikes or surges covered?
- Is a certification required to install or service the product?
- Is the installer able to make equipment modifications, such as custom knockouts or factory-installed wire changes (extend CT wiring)?
- Are there user-serviceable parts?
- What are the most common parts to wear out?
- How are software upgrades performed—remotely by the manufacturer, remotely by the installer? Is there a local method in case there is no on-site internet?

Ongoing (life cycle) costs—Preventive maintenance (PM) is a must for any installed system. Some systems require more care than others, but these basic questions are a good starting point.

- Is PM included or added as a service?
- What are the installer's responsibilities?
- What are the homeowner's responsibilities?
- What are the replacement costs of components as they break or wear out? Who is responsible for them?
- Is there remote system monitoring, data storage, and retrieval? For how long?

- Will the end user receive notifications if the system is not working correctly?
- These systems typically contain different components from different manufacturers. Who is on the hook for each piece, or is there a single support number to call?

For example, a storage system's main components are the solar inverter, the battery inverter, batteries, and, possibly, a generator. Who should the homeowner/installer call if there is an issue with the generator? Or the solar? The homeowner should only have to call the installer. I have been on plenty of service calls with distraught and angry homeowners. They do not enjoy runarounds, and it is up to the installer of the system to be the choke point for all complaints or issues. You've heard the adage "One neck to choke," and as the installer, it should be yours. That doesn't mean you shouldn't bring in the manufacturers when necessary, but typically, manufacturers do not like talking to the end-user for service calls. Most issues require electrical troubleshooting, and most homeowners do not possess that skillset. And for liability reasons, nobody should feel comfortable with a homeowner opening a cabinet, cover, or lid.

It will be up to the installer to determine which manufacturer will assist with the troubleshooting. Piecemealed systems almost always present this challenge. Spoiler alert—it will probably be the individual component manufacturers: talk to the PV inverter company for issues, then the generator manufacturer for their problems, and so on.

2. HOW LONG WILL IT LAST?

This question needs clarification. Are you asking how long the physical unit will last, how long the batteries will last (years), or how long the system will provide power when the grid goes down?

Salespeople usually give the warranty answer: "X-many years or X-many cycles, whichever comes first." Let's look at a standard warranty of 10 years or 7,000 cycles.

- If the unit cycled once per day, the batteries would provide 19 years of service, well past the 10-year stipulation. It is a little generous to imply the system will last that long, unless there is some follow-up in the warranty—for example, an accompanying statement concerning capacity degradation over time. Almost all storage products will reach the yearly proviso before the cycle count.
- Lithium-ion batteries degrade over time, and most warranties include a state-of-health, or degradation, statement. This clause guarantees a percentage of the remaining capacity at the end of the warranty period. Most ESS warranties state 70%. So, if you buy a 10 kWh system, you should have a minimum 7 kWh of remaining capacity at the end of the warranty period.
- Now would be a good time to check if the ESS supports future capacity upgrades.
- Most companies also offer a lifetime throughout rating (50 MWh, 200 MWh, etc.) based on expected performance and cycle count. Some people hold this value in higher regard than the number of warranty cycles. In my opinion, it's all marketing fluff.
- To beat a dead horse, a warranty does not equal reliability. There will be many new entrants to the energy storage market. An energy storage company that started one year ago should provide the same guarantee as a company with 10 or more years of seniority!

Warranty questions also come up with solar inverters and modules. Microinverters and PV modules offer 20-year warranties, although some micros have dropped to 10. String inverters usually have a 10-year warranty with an option to extend to 20. Check with the manufacturer. Regardless, some inverters make it to their warranty period, and some don't. There are always premature failures (see the section on mean time between failures [MTBF]). However, product wear-out will be further down the road. ESSs are no different.

"OK, great, but how long will it last?"

Like every other question in this industry, the answer is the same— it depends! On so many things. Customers may push for an exact amount of time before the equipment needs servicing, but do not fall into that trap. There is no way to predict any of this.

1. Proper equipment location is the single most crucial factor that affects equipment longevity. Indoor or protected conditions are more conducive to longer life cycles. A string inverter installed on the south side of a house will probably experience faster degradation of internal components than one located on another orientation. System parts installed in a garage will likely not suffer the same wear-out time as those installed outdoors. Generally, lithium-ion batteries can tolerate high temperatures, but they prefer cooler temperatures for longevity. I know a guy who keeps his cell phone (and sometimes his laptop) in the hotel mini-fridge when he is traveling.
2. Maintaining recommended clearances is crucial for proper airflow around and into the units. Restricting airflow leads to premature wear-outs due to overheated internal components. Garage installations are the worst offenders. It would be best if you kept the golf clubs and repurposed Amazon boxes away from the equipment.
3. Operating the equipment at the higher ends of rated power will usually require servicing sooner rather than later. Suppose

you have loads that continually require the output power to stay above 80% of the equipment rating. In that case, you might need to rethink the system design. Adding another battery inverter will reduce the strain and promote longevity. If this is out of the customer's budget, reducing loads is the cheapest option. If neither of these options is possible, manage the customer's expectations of equipment longevity.

Now that those variables are out of the way, we can talk about reasonable replacement times. Solar modules will last as long as a mortgage, sometimes even longer. In Golden, Colorado, NREL has modules in their testbed that are over 30 years old and still perform within tolerances and have lost less than the industry-accepted 2% degradation per year. The companies that draft PPAs stipulate module replacement at around 20 to 25 years. These companies are in the business of accounting for every watt-hour and penny spent on the system life cycle.

Only a few solar inverter companies have been around long enough to boast 20 years of field operation. However, those big PPA deals typically account for inverter replacement between 8 and 10 years. These estimates are a safe bet. Battery inverters will probably require the same servicing time frame. I refer you to the battery longevity study from an earlier chapter; however, my gut says lithium-ion batteries will have a 10-year replacement period.

3. HOW LONG CAN I POWER MY HOUSE?

This is a loaded question since many variables influence battery charge and discharge:

- The number of appliances (loads) running affects battery usage.

- More loads mean faster use of the battery, like a phone with multiple apps running in the background.
- These loads put a strain on the battery bank but also on the power electronics.
- A three-ton air conditioner running for eight hours a day typically consumes more battery capacity than the daily use of all other appliances and lighting in the house.

Run time of the appliances (loads):

- The longer the loads run, the more battery capacity they will use.
- An electric stove (2,500 W) that runs for one hour consumes more energy than a refrigerator that runs for 24 hours.
- Loads that use heating elements are the largest consumers. If an electric water heater is on the list, it would be wise to limit those Hollywood showers when the grid is down.[1]

Available solar production:

- Best practices dictate we should not rely on 100% PV production for our estimations since that output rarely occurs.
- Cloudy days equal less production, equal less available PV for direct consumption by the loads, equal less PV available to charge the batteries.
- A PV inverter producing 1,000 W on a cloudy day can only power a microwave, a shop light, or a lot of LED lighting. Unfortunately, there will not be any excess production to charge the batteries while these loads are on.
- Wintertime snow could mean no available PV for several days. Although solar modules can still produce power with about an inch of snow, heavy cloud cover will degrade production to a standstill. The home will be running solely on stored battery

power. A generator is a must for people who live in areas with long periods of heavy snow and cloud cover.

Generator use:

- Always a good idea to have one on-site.
- Charge batteries when there isn't enough sun due to bad weather or storms.
- The generator must be compatible with the ESS recommendations.
- Require routine maintenance.
- Two most common reasons a generator won't start: dead battery or no fuel.

WHOLE-HOME BACKUP

This application is a prevalent request amongst homeowners with the potential to put a tremendous strain on the battery system. Some ESSs or third parties offer breaker-level control with intelligent load management to ease this strain. If the battery system does not support this feature, it is up to the homeowner to manage their loads to prevent a deep battery bank discharge. Not an optimal scenario, even if they assure you they can handle it. Regardless of manufacturer, homeowners not following through on their promises litter most service-call logs.

While it is possible to design a system to back up an entire house, it may not be a good idea. Sometimes the customer isn't right. Even the most carefully planned and expertly sized system could have issues.

- Is the higher-priced system worth the headache of losing the initial profit over repeated service calls?

- Sometimes, the perfectly sized system is out of the homeowner's budget. Shortcuts in system design, drafted under the (mis)understanding that the homeowner will adjust their behavior to compensate, usually lead to repeated service calls.
- Large 240 V loads (anything that requires a two-pole breaker) drain the battery down very quickly—HVAC, ovens, clothes dryers, hot tubs, water/pool pumps, and so on. Best practice dictates strict load-management control.

Whole-home backup applications are not impossible, but they require a more in-depth conversation with the homeowner than a system sized for a sub-panel with specific loads. A customer signature (or three) as acknowledgment of system-design behavior is probably a good idea.

Example:

> I acknowledge, by my signature, that I understand the risks of having large loads powered by my energy storage system and I will manage those loads myself. I further acknowledge I am blowing off this industry professional's warnings and accept these risks because, hey, I have an electrical engineering degree, so how bad can it be? I take full responsibility for my dumbassery if I inadvertently drain the battery bank on the second day of an eight-day outage.

> Sign here: _____

These same considerations are valid even for installations that use a sub-panel chiseled out of the main with those circuits designated for backup use only. However, even these essential-load panels can be susceptible to unforeseen design violations simply because the homeowner might add loads months or even years after the installation is complete. It is not uncommon for a homeowner to add large loads, like a new air conditioner, hot tub, and shop equipment, after installation. Unaccounted for on the original sizing estimate, these loads could put the system into extremis once the grid goes down. I took a call from an installer who was on-site troubleshooting a deeply discharged battery. This system was only three years old and ran flawlessly. As we ran through some comparative troubleshooting steps, he said, "Hey, wait a second. There is a brand-new 30 A breaker on this sub." Since he was the original (and an experienced) installer, he knew this was not on the original design. Can you guess what household item that large breaker belonged to?

Air conditioner. The looming probability of more PG&E-mandated power outages prompted this homeowner to shift this circuit from the main panel to the backup panel. I can't blame them, but the installer did not size the system for this battery-hungry load. The house lost power, and the family was in the dark a few hours later.

SIZING TOOL ESTIMATES

The number-one thing you can do to manage homeowner system-performance expectations is to show them real-world examples.

Sizing tools visually show system performance and the effects of loads and their run times on battery discharge. These estimates will drive home the concepts and relationships of PV production, consumption, and battery behavior. For example:

"You need a system that will provide 30 kWh a day during January, when the power is most likely to go out. Your budget only allows for a 20 kWh system, so we need to talk about alternatives."

Some alternatives are, from most to least expensive:

- Buy more storage.
- Buy more PV.
- Cut down on the loads.

VISUALIZATION OF REAL SYSTEMS

Nothing can drive home the importance of proper system sizing better than real-world examples of system performance. The easiest way to do this is to show homeowners the monitoring portal or app data, preferably of a system in their area and of similar size. I take every opportunity, especially on-site at a new installation, to show off my system. "Excuse me while I whip this out!" Who can argue with the impact this post- or presale display has on someone who has never seen a live, fully operational system?

Considering all these variables, does it still seem likely a sizing tool can recommend an appropriately sized system just by asking you the square footage of your home?

4. HOW DOES IT WORK?

A little knowledge goes a long way for homeowners. The more they understand their system's behavior and normal and abnormal operations, the less likely they are to call you. Your sales and service teams can focus on more important things than answering emails and phone calls like "What does this blinking light mean?"

USER MANUAL

The best defense against unnecessary customer calls is the user or homeowner manual. It should include simple, nontechnical language with a useful glossary with all acronyms spelled out. The user manual does not need too much depth on the solar operation, battery chemistry, charging modes, and other technical specifications or information. It should talk about the main modes of operation, how to change them, and why they would switch from one to another. Some systems use an app to make these changes, while others require a local login to the ESS. The homeowner, and salesperson, should know the answers to these questions.

How Are the Batteries Charged?

- Self-consumption—Only the PV
- Backup—The grid (when available) or solar
- ToU—The grid since peak times are usually late afternoon and evening, when solar does not contribute to battery charging. But an appropriately sized system could rely on solar.

When Is It Necessary to Switch Modes?

A customer with a system running in ToU or self-consumption may want to switch their system to backup mode if a known power outage is imminent: utility-announced outage, winter-storm advisory, severe storm warning, and so on. I also recommend the homeowner play around with the operating modes to see which one best suits their consumption and production behaviors. If the system is in a ToU area, you might be able to open that high-peak window to reduce some grid reliance. My system runs in ToU with the 5:00 to 8:00 p.m. high-peak window extended to 5:00 p.m. to midnight. I forgo grid charging and rely on solar starting at 7:00 a.m., during the off-peak period. I have to keep a close eye on these settings in the winter. If I have a terrible morning of PV production, I may not have a full battery when my peak period begins.

Are There Any Parameters I Need to Change?

ToU mode requires the homeowner to change the peak and off-peak times since those two values will vary depending on the season (typically summer and non-summer rates).

Homeowners should also understand the SoC parameters that limit discharge to preserve capacity if a grid outage occurs (10%, for example).

What Are the Limitations of the Software Platform?

Almost all platforms have a form of remote update. If the internet is down, is there a way to perform local updates?

DIAGRAMS

A picture is worth a thousand words. Visuals serve as reinforcement or as amplifying information. People learn in different ways, and images and diagrams are always useful education-reinforcement aids.

A simple, single-line diagram showing power flow and charging/discharging (like the one in an earlier chapter) can help a homeowner understand the battery system's behavior during grid-up and grid-out conditions. Most manufactures have marketing collateral to help with this imagery. I know solar companies that create marketing materials that are usually better than the manufacturer's.

Videos go a long way in today's world. Most people are looking at screens all day, so you might as well give them something to watch that will benefit both of you. Manufacturer marketing videos can ease service calls if you front-load them with the user manual.

None of this is rocket surgery. Commit to these simple steps and you will increase customer satisfaction and sales while lowering expenses. Most of the hardware for these systems does not offer much room for margin. Soft-cost reduction is the bull's-eye, so aim in early.

"If we hit that bull's-eye, the rest of the dominoes will fall like a house of cards. Checkmate."

—ZAPP BRANNIGAN, CAPTAIN OF THE *NIMBUS*

EXTENDED RANT

SOLAR MYTHS AND MISCONCEPTIONS

"A half-truth is a whole lie."

—Fortune cookie in my General Tso's takeout

I want to begin this section by stating that product variety is critical in this industry because solar rooftops are as varied as their locations and physical appearances. Some rooftops have more shade than others. Some are in hotter climates. Commercial buildings are "flat," while some residential roofs have sharp angles for snow slide. There are advantages and disadvantages to string inverters, microinverters, and optimizers. There is no one-size-fits-all approach. But some technologies would have you believe there is.

Our industry's most commonly accepted truth is that MLEs will produce more energy in shaded conditions. After reading microinverter and optimizer sales and marketing sheets, one would get the impression these things worked *better* in the shade! Everyone has their own "trust the science" confirmation bias. For example, most accredited climate scientists (sorry, Greta) agree that global warming / climate change is real. However, they can't agree on how much the climate is changing, its effects on the planet, or if there is anything we can do to stop it.[1] Conversations concerning the use of MLEs have the same substance. Most reasonable people will agree that MLEs produce more in shaded conditions than a string inverter. But what they cannot agree on is how much and if the overall harvest will offset future servicing costs—if they bother even to ask this question.

I want to begin by saying I am not trying to pick a fight about any of this (my lawyer suggested I preface this section with a statement of liability). I am presenting competing information that corroborates the premise of each claim.

"Why are you beating up on these things so much if you just said product choice is a good thing?"

Two reasons:

1. Misinformation hurts our industry. People deserve to have all the facts so they can make a more informed purchasing decision.
2. MLEs introduce a riskier architecture.

Shuffling through PV inverter sales and marketing materials can feel like a drive down Lombard Street. With complete disregard for common decency, I will dissect some of the more popular claims. I will cite my references, knowing that if 5% of you actually check them, I'm doing good. So let's go: a 20-minute chapter. In and out.

CLAIM 1

If you shade 50% of one module, the entire string will go down by 50%.

String Inverter

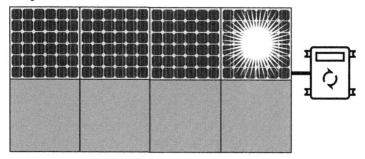

Microinverters

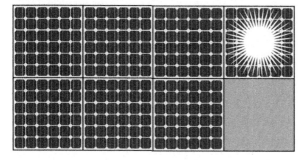

This popular infographic is one of the most frustrating to deconstruct because it takes the most time. This claim is the frail prop for much MLE marketing literature. It is as close to a suggestio falsi as Manny Ramirez explaining how easy it is to hit home runs over the Green Monster at Fenway before handing you a bat with Randy Johnson on the mound.

BACKGROUND

A quick internet search will reveal many hits of this infographic concept—italic emphasis placed by me.

Energy sage: *Do solar panels work in the shade?*[2]

> Though the numbers will vary depending on how much shade the panels are facing, the general rule with clouds and shade is that solar panels will produce about *half as much* energy as they would with direct sunlight…

> For example, if a tree next to your house casts a shadow over one panel in a row on your roof, that entire set of panels will only operate as *efficiently* as the shaded panel.

SunMetrix: *What are the advantages and disadvantages of micro-inverters?*[3]

> You are probably familiar with the old-school Christmas lights that are connected in series where if one bulb fails, it takes the entire string with it. Solar panels attached to a central inverter also have the same vulnerability: one defective panel can stop the entire string from working which will result in a significant or complete loss of electrical supply.

PV Education: *Shading*[4]

> Shading just one cell in a module to half causes the output power of the whole module to *fall to half*. No matter how many cells there are in the string, completely shading one cell causes the output power of the module to fall to *zero*. The lost output power of all the unshaded cells is dissipated in the shaded cell. It is even worse at the system level, where multiple modules are in series to increase

the system voltage to 600 or 1,000 V and shading one cell would affect the *entire module string*.

Notice the universal language that implies one shaded module will drag down the entire array. These descriptions make one wonder how the world ever survived before MLE! You can also see how "panel" is so engrained in the industry vernacular.

Remember this grade-school image of our tongue's taste zones? It is a perfect example of the law of primacy. The first thing you learn is what you commit to memory. It is so important for educators to teach the right thing the first time. It takes seven times longer to unlearn wrong information than to learn it right the first time.

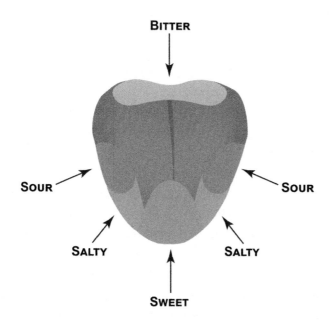

This image is the visualization of a 1901 study by German scientist David P. Hänig. Textbooks immediately latched onto this image,[5] and 120 years later, people still think our tongue is neatly divided into sections solely responsible for those tastes. This diagram has been debunked for decades by biological science, and it doesn't even pass the common-sense test. If our tongue's tip can only taste sweet, why do we use it to taste anything we are trying to sample? OK, back to the argument.

Anyone combing the internet for shading, MLEs, or product information will stumble onto these websites and believe solar modules, arrays, and inverters work in this way. The Law of Primacy will ensure vehement pushback against any competing information. But I'm always up for a challenge, so let the games begin!

REBUTTAL

The "50% shading of one module causes the entire string to drop by 50%" claim:

1. Disregards the use of module bypass diodes
2. Misrepresents the overall effects of shading

Most solar modules use bypass diodes to reroute current away from the affected shaded cells to prevent hot-spot damage. These devices are so common that module manufacturers do not even bring attention to them. Ideally, module manufacturers would use a bypass diode for each cell in the module, ensuring a complete rerouting of shaded cells. To cut costs on these highly commoditized glass rectangles, they typically only use three diodes and group the cells together in three strings. Bypass diodes are fascinating to read about, and there are tons of research papers out there, but most people don't care about reversed bias effects, hot spots, and so on. Installers and homeowners just want the things to work, be safe, and cost as little as possible, and not necessarily in that order.

Let's kill two birds with one stone and take care of rebuttals 1 and 2. A compelling 2019 study entitled *Comprehensive Study of Partially Shaded PV Modules with Overlapping Diodes* explains why shading 50% of one module will not drag the rest of the string or array down by 50%.[6] The peer-reviewed study presents and interprets data from a powerful software-simulation tool. Perhaps, one day, someone will just go out to a ground-mounted array with a tarp, camera, and multimeter to finally put this theory to rest.

"I thought there were already papers where they did just that. They used a tarp and the document fr—"

Whoa, whoa! Hold on there. You're stealing my thunder.

The money shot of this report is Figure 10, which shows mismatch power losses of a partially shaded PV array. There are four array configurations: series, parallel, series-parallel, and total cross tied (TCT). Losses range from 0.24% to 29.3% for the first three configurations. The TCT configuration resulted in the highest losses—46.09%.

"Well, then, there it is: 46.09% is pretty close to 50%. What gives?"

Sure, we can cherry-pick the data that confirms our bias, or we can agree that shading is too complex and has too many variables to be oversimplified by a marketing infographic. Let's use a popular estimation tool to see the effects of shading for ourselves.

Many people don't realize the 8,000 W PV array installed on

their roof will only generate that amount during perfect conditions. There are many ways to predict the output of a solar system using some sophisticated software. Some of these software tools are free, and others are not, but you will find the accuracy of the predictions gets tighter when there is money involved. I am going to use PVWatts again to show the effects of shading. I referenced this tool back in Chapter 13.

PROJECT INFORMATION

Solar resource data: ZIP code 94203

SYSTEM INFO

Modify the inputs below to run the simulation.

DC System Size (kW):	5	ⓘ
Module Type:	Standard ▾	ⓘ
Array Type:	Fixed (open rack) ▾	ⓘ
System Losses (%):	14.08	ⓘ
Tilt (deg):	20	ⓘ
Azimuth (deg):	180	ⓘ

System losses input: Default values

Shading set to 7.6%

Recalculate with shading losses set to 3%

And now for the results. Envelope, please!

drum roll

7.6% shading = 7,502 kWh/year

drum roll

3% shading = 7,877 kWh/year

What does this mean?

drum roll with a cymbal crash

The difference between the two values is 375 kWh per year—the equivalent of a refrigerator-freezer combo. The unshaded system produced 5% more than the shaded system.

Conclusion: shading has a small effect on overall yearly array output.

Let's look at a more sophisticated modeling tool called Helioscope by Folsom Labs and compare results.[7] Unlike the PVWatts estimate, this model compares a string inverter and a microinverter system. Unlike PVWatts, this tool models the effects of the environment on PV inverter behavior. Since banks front the money and demand ROI milestones, robust tools like Helioscope must account for every watt-hour in these projects. I want to express my heartfelt gratitude to Kimandy Lawrence of Lord and Lawrence Engineers, who responded to my LinkedIn request for a Helioscope case study. It is another example of how this industry attracts good people. This software isn't cheap, and I didn't want to spend the money or the time for something I was only going to use once—like when I have to buy a specialized tool to work on my car.

The first project compares shading effects on two different systems using a string inverter and a microinverter. Both projects use the same site location in Florida, the same number of modules, and the same module manufacturer. Both arrays have 10% shading on the same area of the array. The project deliverables did not scale down well enough to include in this book, but here are the highlights. Project available upon request.

One thing of note is the DC to AC ratio for both systems, referred to as load ratio in the reports. The string inverter connects to a smaller array (0.75:1), but the microinverters connect to oversized modules (1.10:1). The first two images represent an unshaded array.

String Inverter System

📊 System Metrics	
Design	No Shade
Module DC Nameplate	2.24 kW
Inverter AC Nameplate	3.00 kW Load Ratio: 0.75
Annual Production	3.401 MWh
Performance Ratio	78.1%
kWh/kWp	1,518.3
Weather Dataset	TMY, 10km Grid (26.65,-80.15), NREL (prospector)
Simulator Version	3882232029-9f87be7113-cb35e05405-8e4bf740e1

Microinverter System

📊 System Metrics	
Design	No Shade
Module DC Nameplate	2.24 kW
Inverter AC Nameplate	2.03 kW Load Ratio: 1.10
Annual Production	3.471 MWh
Performance Ratio	79.7%
kWh/kWp	1,549.4
Weather Dataset	TMY, 10km Grid (26.65,-80.15), NREL (prospector)
Simulator Version	3882232029-9f87be7113-cb35e05405-8e4bf740e1

The unshaded systems show yearly outputs within 2% of each other. The shaded results:

String Inverter System

📊 System Metrics	
Design	With 10% Shade
Module DC Nameplate	2.24 kW
Inverter AC Nameplate	3.00 kW Load Ratio: 0.75
Annual Production	2.838 MWh
Performance Ratio	65.1%
kWh/kWp	1,266.9
Weather Dataset	TMY, 10km Grid (26.65,-80.15), NREL (prospector)
Simulator Version	3882232029-9f87be7113-cb35e05405-8e4bf740e1

Microinverter System

📊 System Metrics	
Design	With 10% Shade
Module DC Nameplate	2.24 kW
Inverter AC Nameplate	2.03 kW Load Ratio: 1.10
Annual Production	3.124 MWh
Performance Ratio	71.7%
kWh/kWp	1,394.6
Weather Dataset	TMY, 10km Grid (26.65,-80.15), NREL (prospector)
Simulator Version	3882232029-9f87be7113-cb35e05405-8e4bf740e1

There is a broader spread of annual production values, and the microinverter system shows a 10.24% increase in production over the string inverter.

The string inverter has a respectable showing but hardly reflects the implied catastrophic behavior shading supposedly has on string inverters. While this simulation software showed a 10% gain in yearly production, real-world examples and other studies show varying results.

In summary, shading part of an array connected to an inverter does not inflict the implied output reduction the image suggests. Now on to the next claim!

CLAIM 2

String inverters cannot handle shade.

REBUTTAL

This misconception does not consider the complex nature of the string-inverter technology and the solar array. Plus, the Helioscope example proved they could.

Shading is a heavily researched topic. One of the earliest (free) papers on the subject was co-written and presented by Henk Oldenkamp at the 2006 EUPVSEC Conference.[8] Studies like these are used by MLE apologists to "prove" string inverters cannot handle shade. Keep the publication date in mind and contrast the inverter technology tested in this study to the cell phone technology we were using in the early 2000s. Granted, I have not had one of these PvP skirmishes in a while, but they occasionally happen.

The backbone of any shade study is the IV curve, a chart that shows the relationship between the array current (I) and voltage (V). Let's take a look at these characteristics that make up an IV curve in action.

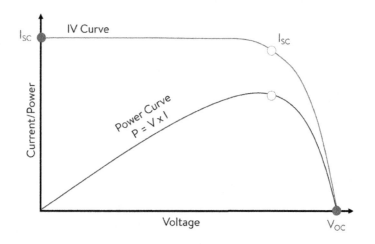

This plot is calculated using some serious math or by direct measurement using sophisticated test equipment. There are no values on this IV curve since I want to focus on the shape first. There are many external forces (nonmanufacturing influences) that affect the IV curve of a module:

- The amount of available sunlight (irradiance)
- Temperature
- Geographic location
- Orientation
- Angle (roof pitch)
- Shade

I do not want to turn this into a dissertation on IV curves since the study explains it in sufficient detail for anyone who wants to dig a little deeper. The first thing to understand is what happens to the IV curve when there is shading. Instead of the typical shifted bell-curve shape, the IV curve will develop "maxima," or humps.

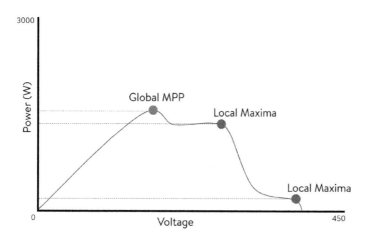

Depending on the severity of the shade, the IV curve could have multiple humps; however, there is always one that will still produce that maximum amount of available power from the array (maximum power point, or MPP). This hump is called "global maxima," or "global peak." The other humps are called "local maxima" and could fool the inverter if it stops at these locations during the tracking process.

"What do you mean by this 'inverter tracking'? I don't get it."

I guess we need to dive a little into how PV inverters find that global peak.

"Oh, sweet! More technical lecturing. So glad I asked!"

Every PV inverter has an MPPT algorithm to help generate the maximum amount of power possible as the day progresses. Once the PV inverter is satisfied with its tracking on the IV curve, it will adjust its track off of and back onto the MPP at regular intervals throughout the day, month, and years of its lifetime. Some PV inverters have sophisticated and highly accurate MPP tracking software that ensures they will always track the global peak. Even in shaded conditions!

Shade-tolerant trackers have been around for a while, but not every PV inverter has this technology. This specialized MPPT will sweep the entire IV curve at regular intervals, instead of just bouncing around the global peak. Think of a radio scanner that is going back and forth, looking for a signal. As soon as it finds the strongest one, it locks on.

String-inverter manufacturer videos demonstrate this shade-tolerant MPPT very well. They all show the IV curve changing as the irradiance changes throughout the day, including with the introduction of shade, and demonstrate how the MPPT follows right along. To reiterate, while all inverters have some sort of MPPT, not all have the shade-tolerant variants. I wonder if shade studies include those types? Regardless, these inverters reduce the effects of shading and debunk the original claim against them.

"But…but…I found a study that corroborates the notion that string inverters cannot deal with shade. This study proves it!"

shoves fistful of stapled, crumpled documents into my face

Oldenkamp's study reveals some interesting PV inverter behavior. Some inverters can find the global, some can't, and some don't even try. I must point out that inverter technology in 2003 could very well already be in a Smithsonian exhibit. Unfortunately, nobody connected to the study replied to my requests to use the graphics, probably because the authors' email addresses no longer work. Refer to the endnotes if you would like to read the study for yourself.

FIRST TEST

None of the 13 PV inverters found the global peak on a complex IV curve like the one above. Most of them tracked the higher local peak, and a few stopped at the lower local peak.

SECOND TEST

A less complicated IV curve with only two humps produced better results, but only about half of the inverters found the global peak, and the other half got stuck on the local peak.

Seldom identified in many of these independent tests and studies are the makes and models of tested components—probably because of our society's litigious nature. However, based on this study's age, I would bet a case of beer that none of them included a PV inverter with shade-tolerant MPPT. I wonder if the modules even had bypass diodes. The study does not specify either. Regardless, it appears things are not going well in this study for the string inverters.

Wait, someone just handed me a note.

Oh, did I mention these inverter MPPT algorithms were tested by shutting the inverters off, shading the array, and then turning them back on? They had to turn the inverter off to trick it!

Wait, someone just handed me another note. I also forgot to mention that tracking off on the local peaks only resulted in a 1% to 2% reduction in yield (see Claim 1 Rebuttal).

Fun fact: most peer-reviewed studies conclude PV inverters track the global peak with no problem under normal operating conditions.

So now we bring it all back around to the basic premise of this claim that string inverters cannot handle shade, which handshakes with Claim 1. This claim:

1. Disregards the complex nature of shading on the individual components' behavior or the PV system
2. Misrepresents the overall effects of shading. While the shading might degrade the power output at a particular time of day, the degradation does not continue throughout the day, that is, a 13% loss of power was only for that time frame.
3. Does not account for the complex technology used in modern string inverters to reduce shading effects

But wait, there's more! A bonus track:

4. The claim conflicts with manufacturer-sponsored studies that show the influence of shade has a minimal effect on the long-term overall energy yields, regardless of tested inverter technology. More on this later.

CLAIM 3

MLEs will produce 25% (or more) more power in shaded conditions than string inverters.

REBUTTAL

False. False. False. This claim is rooted in a misinterpretation of MLE marketing literature from years ago, which stated optimizers would *recover* 25% of the lost power from the shaded array. Recovering 25% is *not* the same as producing 25% more.

CLAIM 4

MLEs reduce soiling losses, increase system efficiency, and lower system downtime (redundancy) if one fails.

REBUTTAL

If a string inverter fails, the solar system production is reduced to zero until the inverter is repaired or replaced. If one microinverter or optimizer malfunctions, the rest of the system will continue to operate. Some microinverters connect to two, three, or even four modules instead of to a single module. This concept reduces the number of MLEs on the roof and reduces the future risk for installers when replacing them. I recall talking to an SMA engineer about this, and she said, "Interesting. But if reducing redundancy is such an advantage, then why not just use a single string inverter with a shade-mitigation MPPT?" *Bam!*

Shouted from the rooftops, system redundancy is purported to be and blindly accepted as a distinct advantage for any module-level device. And I agree. However, that redundancy breaks down at some point. Hold that thought for a few more pages.

CLAIM 5

Microinverters have a 300-year MTBF, while a standard string inverter only has 12 to 15 years. Thus, microinverters are more reliable.

REBUTTAL

I am not fond of this claim. It seems sleazy to me. MTBF is an engineering value of reliability used to guide the manufacturer's recommendations for routine inspections, replacement, or repair.

Originally coined and calculated for the US military, MTBF has a few flaws, making it a dismissive claim for our industry.

1. It assumes failure rates are constant (which they are not).
2. It is a fancy method for expressing failure rates (MTBF is the inverse of failure rate).
3. It is a probability of failure, not a duration. Three hundred years between failures? Really?
4. It assumes no wear-outs.
5. It assumes no infant mortality.
6. Failures will occur randomly.

Despite all these shortcomings, some still use it as an advantage over string inverters. For the record, string inverters have an industry-accepted MTBF of 150 years; however, 10 to 15 years seems to be the likely time frame for most of them to wear out.

Fronius published a technical white paper titled "Sustainability for the PV Industry: Field Service" and corroborated the 150-year value.[9]

"Oh, a string-inverter company stating their string inverter MTBF is 100 to 200 years. How convenient."

OK, how about a microinverter company stating that MTBF is misused? Enecsys CEO Paul Engle said, "There is a common misunderstanding that a product with a high MTBF will last longer than a similar product with a lower MTBF."[10]

Engle goes on to deconstruct this misunderstanding in a very compelling way. In my opinion, throw out any MTBF points when talking about the advantages of MLE.

Let's look at three modern-day studies that compare MLE and

string-inverter production in shaded conditions. Maybe we can finally get a ruling on this once and for all!

MANUFACTURER-SPONSORED SHADE STUDIES

This section will briefly compare three different manufacturer-sponsored shade studies. These studies, once and for all, prove that X outperforms Y and Z. How exciting!

STUDY #1—SOLAREDGE-SPONSORED STUDY

The first study was conducted by PV Evolutions Lab (PVEL) in 2013 using a shade-study standard created by the NREL.[11] The study included applying varying amounts of black screen to the arrays to create light-, medium-, and heavy-shaded conditions. The SolarEdge system produced 1.9%, 5%, and 8.4% more energy than the SMA string inverter. It accomplished these production increases by recovering a portion of the lost energy the string inverter suffered because of the shading—28.3%, 21.9%, and 24.2%. The keyword here is "recovered."

Unfortunately, this study's unintended casualty is the misinterpretation that a DC-optimized system produces 25% more energy than a string inverter. This misinterpretation was then casually applied to microinverters. This was the birth of the "MLEs produce 25% (or more) more energy than string inverters" myth.

The optimized system also outperformed the microinverters.

STUDY #2—ENPHASE-SPONSORED STUDY

PV Evolution Labs also conducted a study that compared the outputs of a microinverter to a string inverter under shaded conditions.[12] One thing that stood out is that the array for the 3,000 W string inverter was undersized compared to the microinverter's DC to AC ratio. The microinverter system consisted of twelve 240 W modules connected to a 215 W microinverter. This setup had a DC to AC ratio of 1.11 (oversized). The string system used the same 12 modules to reach a DC wattage of 2,880 W. However, the DC to AC ratio was only 0.96 (undersized). I can see why the testing would use the same array size for each system, but it is not fair, in my opinion. Although both systems share an array size of 2,880 W, they will behave much differently.

Any PC gamers out there? Here's a relatable analogy. This test gives the microinverter system an unfair advantage because it is essentially an overclocked design. A much larger solar module connects to the microinverter. Therefore, the micros will produce a much higher output level (and for a longer time during the day) than the string inverter connected to an array that doesn't even match its nominal AC output. It is the same ratio conversation from the Helioscope case study.

There were some weighting factors that PV Evolution Labs used to try and even the odds, but, interestingly enough, even the "overclocking" didn't matter. As with the previous study comparison, the microinverter harvest beat the string inverter by a paltry 1.2% in a study that ran for almost a year.

Author's note: I find it interesting that a testing lab used the NREL testing standards to compare yields for the same PV inverter technologies in two different tests and came up with two different results.[13]

STUDY #3—SMA-SPONSORED STUDY

This 2019 third-party study performed by the University of Southern Denmark compared three different systems[14]: an SMA string inverter, an SMA string inverter paired with Tigo DC optimizers, and a SolarEdge DC-optimized system. The study ran for three months and went into greater detail on MPPT and IV-curve behavior than the previous two studies.

The stand-alone SMA inverter with advanced shade-tolerant MPPT outproduced both the SMA inverter with Tigo optimizers and the SolarEdge optimized system in partially shaded conditions, but only by a mere 0.2% to 0.4%. The results were fascinating but not surprising for two reasons:

1. They conclude, once again, that shade has a small effect on overall inverter output.
2. I have seen the SMA MPPT algorithm in action using the same scientific equipment NREL uses for their testing.

A Fronius 2016 study, "A Yield Comparison between DC Optimized Systems and Conventional PV Systems Using Fronius Inverters,"[15] also concluded that string inverters using DC optimizers produce less power than a string-inverter system using only advanced MPPT. Optimizers consume more energy than they make under certain conditions.

Is it so hard to believe software can come close to and perhaps even surpass hardware in this day and age? For the record, I am not anti-microinverter or anti-optimizer. As I have repeatedly said, there are genuine and clear-cut reasons to use them. I am anti-sales-and-marketing bologna. There is no place for it, and it hurts our industry.

"So why do companies and websites keep doing it?"

A better and less litigious question is why our industry continues to let them get away with it. If I were a module or string-inverter manufacturer, I would be pissed if someone came along and made a sweeping and overgeneralized statement that cast doubt on all the research and development, technology, and resources I put into my product. But I'm not the only one questioning the MLE claims. Blair Reynolds, solar and storage product manager for SMA America, tagged me in his LinkedIn post, "The Top 10 Most Common Myths about Traditional DC Power Optimizers."[16] The article echoes many of my observations but touched on one myth in particular that merits a plug.

Myth: Power optimizers (or microinverters) are required for code compliance.

Nein! While they might be the easiest way to comply with rapid shutdown in some circumstances, they are not required.

"OK, hold up. You just said they are an easy way to comply with this requirement. So if they comply, why all the fuss?"

As I alluded earlier, MLEs have a downside I am not sure many installation companies consider. Are those small gains in production worth the extra risk? Let's explore that question.

THE HIGH COST OF MODULE-LEVEL ELECTRONICS

GREG'S AXIOM #1

Regardless of make and model, every solar inverter will suffer wear-out, even if it does not fail.

GREG'S AXIOM #2

Roof equals risk.

A fictitious solar company called Willie's Solar, Plumbing, BBQ, Pest Control, and Diamonds, Inc., shifted its business model to using MLE exclusively. Willie's has been on a tear for the last five years and reached 1 MW of installed capacity—4,000 units across 200 installations. Not too shabby. While the MLE advertised *failure rate* is roughly 1 in 2,000, the probability of *wear-out* is 4,000 out of 4,000. We don't know when that will be, but there are some certainties:

- The MLE will not all wear out at the same time.
- The MLE will not all wear out at the same location at the same time.
- A truck roll is required for the removal and replacement of each wear-out.

These are inevitable truths for any MLE installation. The most critical question you need to ask yourself as a solar installation company owner is this: what is our plan for the future removal and replacement of our installed microinverters and optimizers?

I have heard many maintenance plans over the years, ranging from responsible to reprehensible. Some installers send crews out a few days a month to ride around and replace bad components, while others just let the failures stack up until the system output is noticeable enough for the homeowner to complain. Here are the most common counterarguments, or "yabuts":

"Yeah, but I get a $150 truck-roll compensation from the manufacturer."

The average truck-roll cost is between $300 and $600 (GTM numbers from 2105[17]). Let's call it $400, which seems to be a fair assumption and corroborated by the installers who attend my classes.

Four hundred dollars minus $150 equals $250 coming out of your pocket every time you roll a truck to replace a bad MLE. It also puts the crew at higher risk since the removal and replacement requires roof work. That's a sobering thought for any business owner. But this argument hinges on the fact that you will even get a reimbursement.

Achtung! Some MLE companies have phased this service agreement out of their warranties. D'oh!

"Yeah, but the monitoring system will tell me when the panels go bad."

Sure, modules go bad occasionally, but are you running right out to the site to swap it out? I doubt it. The monitoring will allow you to find failed or worn-out MLE. But I bet dollars to donuts you don't rush right out and swap out that bad MLE.

"Yeah, but there is more redundancy in an MLE system. If one fails, the entire system will continue producing power. If the string inverter fails, the whole system stops producing."

Right, and I am a big fan of redundancy, but when does the advantage break down? Submariners live by the "Have a backup for the backup" mantra, and so does the engineering world. The small chance of a failure in a redundant system weighed against either a riskier roof swap-out for the MLE (multiple times during system lifetime) or a ground-level replacement for a string inverter (once in 20 years?) seems to be a no-brainer. It is not the failures I am concerned about but rather the wear-outs.

"Yeah, but there are some legit yabuts you haven't mentioned. Sure, I agree with the premise that micros and optimizers are riskier. I don't buy the 50% string graphic, yadda, yadda. But you don't know what it's like to be on-site and not be able to reach technical support after being on a roof all day. It is frustrating. Sure, Brand X may be the best, but their service line won't call me back if the house is on fire. And that's about when Brand Y might call me back. At least Brand Z picks up the phone. To me, that's worth the extra risk or hassle of a slightly higher failure rate.

"And those things are easy to use and install. That's worth money to me. Furthermore, my business model is a sound one that I have develop—"

OK, OK, I get it. When you say it like that, it makes me sound like a jerk. But I agree! Service is an essential part of the equation that I have not touched on. Bad customer service doesn't do anyone any good—not the installer, the manufacturer, or the homeowner who is eight times more likely to talk about a bad experience than a good one. For some installers, the risk of future MLE wear-outs is acceptable compared to a manufacturer that does not offer support.

GREG'S AXIOM #3

There is plenty of room in this industry for everyone.

Many MLE companies did wise up and change their marketing literature, and I applaud them for it. The industry is a lot better off because of its integrity. However, as I've demonstrated, there is still wrong information lingering around these initial misconceptions. Sadly, there probably always will be. But now you have the facts to refute these claims and make a more informed decision on which technology you wish to use for your business.

To summarize:

- People often overestimate the effects of shading.
- Some MLE performance infographics require too many footnotes to be accurate.
- Most string inverters use sophisticated MPPT to reduce shading.
- Decide if you like spending time on the roof or at ground level.
- Have an MLE removal and replacement plan.
- There's plenty of room in this industry for everyone!

CLOSING THOUGHTS

You made it! Time for celebration since this material, although crucial for a successful business model, can be as dry as a box of salted crackers. You should better understand how the solar and storage system components work together and how to size these systems for the right application. The jargon and terminology of spec sheets should be less intimidating. Even if you don't agree with some of my rants, there are still some lessons to be learned. If you started this book as a layman, you should be well on your way to becoming the life of the party if someone broaches the subject of solar and storage. Speaking of broaching, I feel like I'm back on the boat after a long underway, and we have surfaced for our final port ingress—relieved, excited, and anxious. Writing this book has been a fantastic journey, and it is my sincerest wish that you have some fruitful takeaways. I hope you feel the same way!

I admonish the same call to action as I do when I come to the end of a training day. Remember the *Starship Troopers* scene when they landed on the alien bug planet to take out the hive mind? On the ride down, the man in charge tells the squad, "Remember your training, and you will come back alive!" Although not as dire as fighting wave after wave of angry aliens, this training will keep you from talking to an angry Robert Beckett. Nothing beats a happy customer, and if they are happy, you are happy.

There is a lot of opportunity in this industry, and it is there for the taking. This industry attracts good people, and there is more than enough room for everyone. We are essential workers. Growing the renewable-energy industry is not a side quest. It is the main campaign mission that we must fight through to the final boss.

If you are at a solar conference or exhibition, I am always ready to talk shop or swap sea stories over a few beers. The first round is on me!

GRATITUDES

Thank you to SMA America for hiring me back in 2008. Transitioning from the military in 2007 was tough. I will always be grateful for the chance you took with me. Dave W., you showed me the fun side of sales. Ben C., you kept me sane when things got crazy. Jorg F., thank you for inviting me into your quaint German home and taking me to see a real German football game! I still have that Deutschland hat. Your incredible depth of knowledge inspired me to dig into the industry's storage side. My time at SMA was one of the best in this industry. You are all awesome! Matt, this one's for you, buddy: https://tinyurl.com/uk67ymr8.

My Solar Energy International (SEI) extended family are some of the best people I have had the pleasure to train and socialize with over the years. Kris and Kathy, I will always hold a special place in my heart for Paonia and your hospitality (especially the care package of elk steaks!). I genuinely enjoy my time there and appreciate the opportunity to present my knowledge to your classes. As I have said, this industry attracts good people, and eventually, they make it to Paonia.

Thank you, Justine Sanchez, my SEI sister, for the help you have given me over the years and for your guidance on this book.

Ryan Mayfield, I thank you for your devotion to this industry and for sharing the incredible NEC knowledge you possess. I've never seen you lose your cool, even after hearing the same questions over and over and over. I appreciate the assists over the years.

Ezra, thank you so much for sharing your wealth of knowledge and your words of encouragement over the years. Living off-grid for 50 years is an incredible feat, and the world is a better place because of people like you. I still tell people about the turkey incident from the first NABCEP conference in Upstate New York.

Speaking of NABCEP, I can't thank that organization enough for their hard work and the annual continuing-education conferences. It's good street cred in a growing industry that continues to grab people from all backgrounds and walks of life. The learning really happens after the days-long sessions during happy hour. Sue, you are a rock star, and you keep those events running smoothly!

A huge thank-you to Nicholas Carter, PhD, owner of npc Solar, who reviewed the manuscript and highlighted all the snide comments, vulgarity, and incorrect calculations for correction.

Steve M., former sonarman and friend, I am not sure where I would be today if it weren't for your support during those years on the *L.A.* and at NSTCP. It was tough for me personally and professionally, and you didn't give up on me. I learned how to deal with the mini-mes and the senior leadership who couldn't find their asses with a flashlight while standing over a mirror. We learned how to laugh during dark times, and we took care of our own. We outlasted them all, and it wasn't even close to Halloween.

A heartfelt THANK-YOU!! to the Scribe Media team who helped me get through this book—Rachael, Miles, and Rikki. You guys are amazing and eased my nerves as we moved from one phase of the process to the next. Special thanks to my publishing manager, Katie, who put up with my barrage of questions and contrariness.

ABOUT THE AUTHOR

 Greg began his military career in 1987 as a submarine sonar technician. He spent most of his time in Pearl Harbor, Hawaii, when he wasn't punching holes in the ocean for weeks or months at a time. After retiring in 2007, Greg began his renewable-energy career in 2008 with SMA America as the senior technical trainer. Since solar is just one letter away from sonar, he figured it would be a smooth transition. After seven and a half years, Greg left SMA and went to sonnen, Inc., an energy storage manufacturer, and helped launch this new startup in the United States.

Greg is well known throughout the industry as a passionate and engaging presenter with a down-to-earth teaching style. He is a published and recognized subject-matter expert routinely sought out by industry professionals for his knowledge and advice.

While most people worked from home in 2020, Greg took advantage of 0% travel by honing his barbecue skills—which is the only thing he enjoys more than educating. Whiskey is a close second. Greg has lived in the Sacramento area since 2007. He hunts, plays guitar, and is an avid gamer.

NOTES

CHAPTER 1

1   Solar Energy International, "Online Solar Training & Renewable Energy Courses," https://solarenergytraining.org/.

2   Wood Mackenzie Power & Renewables and Solar Energy Industries Association, *Solar Market Insight Report 2020 Year in Review* (Solar Energy Industries Association, 2021).

CHAPTER 2

1   "Electricity Rates by State," Electric Choice, accessed December 2020, https://www.electricchoice.com/electricity-prices-by-state/.

CHAPTER 3

1   STC definition and criteria: https://www.pvsyst.com/help/stc.htm.

2   A. Kimber et al., "Improved Test Method to Verify the Power Rating of a Photovoltaic (PV) Project" (paper presented at 34th IEEE Photovoltaic Specialists Conference [PVSC], Philadelphia, PA, 2009, pp. 000316–000321), doi: 10.1109/PVSC.2009.5411670.

3   Topaz Solar Farm is a First Solar project: http://www.firstsolar.com/Resources/Projects/Topaz-Solar-Farm.

CHAPTER 4

1   Wood Mackenzie Power & Renewables and Solar Energy Industries Association, *Solar Market Insight Report 2020 Year in Review* (Solar Energy Industries Association, 2021).

2   "Average Price of Solar Crystalline Silicon Photovoltaic Modules in the United States from 2008 to 2016," Statista, accessed March 20, 2021, https://www.statista.com/statistics/790521/standard-prices-for-solar-pv-modules/.

3   Oldenkamp's website has a lot of useful information that dates back to solar's early days: http://oke-services.nl/.

4   Mike Munsell, "Who Leads the US Residential Inverter Market?" GTM, October 9, 2014, https://www.greentechmedia.com/articles/read/who-is-leading-the-u-s-residential-inverter-market.

5   I am not implying people who use microinverters are dummies. Micros offered a much easier solution than the market was used to.

6   Big shout-out to Jeff for this documentary! *Solar Roots: The Pioneers of PV*, produced by Jeff Spies and Jason Vetterli (2017; undistributed), https://www.solar-roots.com/.

CHAPTER 5

1   "US Solar Market Sets New Record, Installing 7.3 GW of Solar PV in 2015," *SEIA*, February 19, 2016, https://www.seia.org/news/us-solar-market-sets-new-record-installing-73-gw-solar-pv-2015.

2   "Behind-the-Meter Energy Storage: An Emotional or Financial Decision?" Wood MacKenzie, June 19, 2018, https://www.woodmac.com/our-expertise/capabilities/power-and-renewables/extracting-value-from-energy-storage-participation-in-energy-markets-can-boost-customer-adoption/.

3   Term of endearment.

4   "Phase Out Peakers: Replacing Polluting Urban Power Plants with Renewables and Battery Storage," Clean Energy Group, May 7, 2020, https://www.cleanegroup.org/ceg-projects/phase-out-peakers/.

5   Robert Walton, "Project of the Year: Soleil Lofts solar+storage development," *Utility Dive*, December 9, 2019, https://www.utilitydive.com/news/project-soleil-sonnen-pacificorp-rmp-batteries-solar-dive-awards/566230/.

6   Bella Peacock, "'Disrupters' Tip Australia to Become World's Biggest VPP in 2021," *PV Magazine Australia*, October 28, 2020, https://www.pv-magazine-australia.com/2020/10/28/disrupters-tip-australia-to-become-worlds-biggest-vpp-in-2021/.

CHAPTER 6

1   Anuradha Varanasi, "For Developing Countries, More Solar Power—and Maybe More Lead?" *Undark*, June 3, 2020, https://undark.org/2020/06/03/india-solar-power-lead/.

2   US Energy Information Administration, "Battery Storage in the United States: An Update on Market Trends," US Department of Energy, July 2020, https://www.eia.gov/analysis/studies/electricity/batterystorage/pdf/battery_storage.pdf.

CHAPTER 7

1   Yasin Emre Durmus et al., "Side by Side Battery Technologies with Lithium-Ion Based Batteries," *Advanced Energy Materials* 10, no. 24 (2020), https://doi.org/10.1002/aenm.202000089.

2   Tyler Or et al., "Recycling of Mixed Cathode Lithium-Ion Batteries for Electric Vehicles: Current Status and Future Outlook," *Carbon Energy* 2, no. 1 (2020): 6–43, https://doi.org/10.1002/cey2.29.

3   Hector Beltran, Pablo Ayuso, and Emilio Pérez, "Lifetime Expectancy of Li-Ion Batteries Used for Residential Solar Storage," *Energies* 13, no. 3 (2020): 568, https://doi.org/10.3390/en13030568.

4   Andreas Dinger et al., "Batteries for Electric Cars: Challenges, Opportunities, and the Outlook to 2020," *BCG Focus*, (2010): 3, https://image-src.bcg.com/Images/BCG_Batteries_Electric_Cars_Dec_2009_small_tcm9-120284.pdf.

5   Peter Maloney, "Electric Vehicle and Stationary Storage Batteries Begin to Diverge as Performance Priorities Evolve," *Utility Dive*, August 1, 2018, https://www.utilitydive.com/news/batteries-for-electric-vehicles-and-stationary-storage-are-showing-signs-of/528848/.

6   Andy Colthorpe, "Choice of Lithium Iron Phosphate Not a 'Silver Bullet Solution' for Safety," *Energy Storage News*, August 21, 2020, https://www.energy-storage.news/news/lfp-vs-nmc-not-a-silver-bullet-solution-for-safety.

7   Beth Miller, "Materials in Lithium-Ion Batteries May Be Recycled for Reuse," Phys.org, September 15, 2020, https://phys.org/news/2020-09-materials-lithium-ion-batteries-recycled-reuse.html.

8   Marian Willuhn, "The Issues with Lithium-Ion Battery Recycling—and How to Fix Them," *PV Magazine*, October 28, 2020, https://www.pv-magazine.com/2020/10/28/the-issues-with-lithium-ion-battery-recycling-and-how-to-fix-them/.

CHAPTER 9

1   "The Mayfield Design Tool," Mayfield Renewables, https://www.mayfield.energy/design-tool-request.

2   Tesla's online system calculator, https://www.tesla.com/powerwall/design.

3   *Ruud Choice Series Air Conditioners* (Fort Smith, AR: Ruud), https://s3.amazonaws.com/WebPartners/ProductDocuments/2E208432-C83F-4079-9811-7907E7624A7F.pdf.

4   "NEMA Code Letters for Locked-Rotor KVA," Joliet Technologies, accessed March 20, 2021, https://joliettech.com/information/easa-electrical-engineering-handbook/nema-code-letters-for-ac-motors-easa-electrical-engineering-handbook/.

CHAPTER 10

1   "Renewable Energy Battery Sizing Calculator," Trojan Battery Company, accessed March 20, 2021, http://www.batterysizingcalculator.com/.

2   US Department of Energy, "Estimating Appliance and Home Electronic Energy Use," Energy.gov, accessed March 20, 2021, https://www.energy.gov/energysaver/save-electricity-and-fuel/appliances-and-electronics/estimating-appliance-and-home.

3   "Flip Your Fridge Calculator," Energy Star, accessed March 20, 2021, https://www.energystar.gov/products/appliances/refrigerators/flip-your-fridge. (Why does fridge have a "d" but refrigerator doesn't?)

4   "See How SunVault Can Power Your Home," SunPower, accessed March 20, 2021, https://us.sunpower.com/home-solar/solar-battery-storage#.

5   "SMA Sunny Design," SMA, accessed March 20, 2021, https://www.sunnydesignweb.com/.

6   Blue Planet Energy, https://blueplanetenergy.com/.

7   "Home Battery," LG Energy Solution, https://www.lgessbattery.com/eu/main/main.lg.

8   "Powerwall," Tesla, https://www.tesla.com/powerwall.

9   "PWRcell Solar Batter + Storage System Brochure," Generac, accessed March 20, 2021, https://www.generac.com/generaccorporate/media/library/content/clean%20energy/pwrcell_consumer_brochure.pdf?ext=.pdf.

CHAPTER 11

1   Outback Power, https://outbackpower.com/products/energy-storage/energy-storage.

2   Schneider Electric, https://solar.schneider-electric.com/solutions/residential/.

3   Sol-Ark, https://www.sol-ark.com/.

4   "The Leviton Load Center," Leviton, accessed March 20, 2021, https://www.leviton.com/en/
    products/residential/load-centers.

5   Lumin, https://www.luminsmart.com/.

6   Span, https://www.span.io/.

7   0.8 is a system-efficiency value. Although there are system-efficiency-loss calculators, 80% seems
    to be an industry-accepted standard that is close enough for this exercise.

8   "StorEdge Solutions," SolarEdge, accessed March 20, 2021, https://www.solaredge.com/us/
    StorEdge-solutions.

CHAPTER 12

1   "#WellActually, Americans Say Customer Service Is Better than Ever," BusinessWire,
    December 15, 2017, https://www.businesswire.com/news/home/20171215005416/en/
    WellActually-Americans-Say-Customer-Service-is-Better-Than-Ever.

2   Steven D. Levitt and Stephen J. Dubner, *Freakonomics: A Rogue Economist Explores the Hidden
    Side of Everything* (New York: William Morrow, 2005).

CHAPTER 13

1   "Solar Inverters," SMA, accessed March 20, 2021, https://www.sma-america.com/products/
    solarinverters.html.

2   National Renewable Energy Laboratory, "PVWatts Calculator," NREL, https://pvwatts.nrel.
    gov/. NREL is a national laboratory of the US Department of Energy, Office of Energy
    Efficiency and Renewable Energy, operated by the Alliance for Sustainable Energy, LLC.

3   Hawaiian word for "buttocks."

4   Rocky Mountain Institute, Homer Energy, and Cohnreznick Think Energy, *The Economics of
    Grid Defection*, (Boulder: Rocky Mountain Institute, 2014), https://rmi.org/insight/economics-
    grid-defection/. (I'm still not convinced.)

5   "How Much Energy Does Your Smartphone Use?" The Whiz Cells, September 29, 2016,
    https://www.thewhizcells.com/smartphone-energy-use/.

CHAPTER 14

1   Normal showering procedure to save precious water: turn on the shower and get wet, turn off the water, soap up, turn on the shower and rinse, repeat for shampoo. Someone who let the water run while they stood under the spray for a lengthy amount of time is said to have taken a Hollywood shower. Hollywood showers are a luxury aboard submarines, and most people caught indulging in one would exit the shower to find their towel was missing as they tried to find the light switch to the head (Navy word for bathroom).

EXTENDED RANT

1   "Only 50% of Scientists Blame Mankind for Climate Change in New Study," Media Research Center, accessed March 20, 2021, https://www.mrc.org/articles/only-50-scientists-blame-mankind-climate-change-new-study.

2   Jacob Marsh, "Do Solar Panels Work in the Shade?" Energy Sage, June 20, 2017, https://news.energysage.com/solar-panels-work-shade/.

3   Shen Ge, "What Are the Advantages and Disadvantages of Micro-Inverters?" SunMetrix, September 16, 2014, https://sunmetrix.com/what-are-the-advantages-and-disadvantages-of-micro-inverters/.

4   "Shading," PV Education, accessed March 20, 2021, https://www.pveducation.org/ko/%ED%83%9C%EC%96%91%EA%B4%91/7-modules-and-arrays/shading.

5   Stephen D. Munger, "The Taste Map of the Tongue You Learned in School Is All Wrong," *Smithsonian Magazine*, May 23, 2017, https://www.smithsonianmag.com/science-nature/neat-and-tidy-map-tastes-tongue-you-learned-school-all-wrong-180963407/.

6   J.C. Teo, "Effects of Bypass Diode Configurations to the Maximum Power of Photovoltaic Module," *International Journal of Smart Grid and Clean Energy* 6, no. 4 (2017): 225–232, accessed March 20, 2021, http://www.ijsgce.com/uploadfile/2017/1030/20171030042644575.pdf.

7   "Helioscope," Folsom Labs, accessed March 20, 2021, https://www.helioscope.com/.

8   Roland Bruendlinger, Benoît Bletterie, Matthias Milde, and Henk Oldenkamp, "Maximum Power Point Tracking Performance under Partially Shaded PV Array Conditions" (paper presented at EUPVSEC Conference, Dresden, Germany, 2006), http://oke-services.com/downloads/200609_other_paperepvsec21.pdf.

9   Brian Lydic, "Sustainability for the PV Industry: Field Service," Fronius, March 27, 2014, https://www.fronius.com/~/downloads/Solar%20Energy/Technical%20Articles/SE_TEA_PV_Sustainability_Study_EN_US.pdf.

10 Paul Engle, "MTBF and Reliability—A Misunderstood Relationship in Solar PV," *Renewable Energy World*, September 24, 2009, https://www.renewableenergyworld.com/solar/mtbf-and-reliability-a-misunderstood-relationship-in-solar-pv/.

11 "Performance of PV Technologies under Shaded Conditions," Solar Edge, April 2020, https://www.solaredge.com/sites/default/files/performance_of_pv_topologies_under_shaded_conditions.pdf.

12 Jason Forrest and Nir Jacobson, "Enphase Energy," PV Evolution Labs, February 18, 2014, https://enphase.com/sites/default/files/Study_Comparing_Enphase_and_SMA.pdf.

13 Chris Deline, Jenya Meydbray, Matt Donovan, and Jason Forrest, "Photovoltaic Shading Testbed for Module-Level Power Electronics," NREL, May 2012, https://www.nrel.gov/docs/fy12osti/54876.pdf.

14 W. Toke Franke, "The Impact of Optimizers for PV-Modules," University of Southern Denmark, May 2019, https://www.sdu.dk/-/media/files/om_sdu/centre/cie/optimizer+for+pv+modules+ver11_final.pdf.

15 "A Yield Comparison between DC Optimized Systems and Conventional PV Systems Using Fronius Inverters," Fronius, November 2016, https://www.fronius.com/-/downloads/Solar%20Energy/Whitepaper/SE_WP_A_yield_comparison_between_DC_optimised_systems_and_conventional_PV_systems_using_Fronius_inverters_EN_AU.pdf.

16 Blair Reynolds, "The Top 10 Most Common Myths about Traditional DC Power Optimizers," SMA (blog), February 2, 2021, https://www.sma-sunny.com/us/the-top-10-most-common-myths-about-traditional-dc-power-optimizers-part-1-of-3/.

17 Jeff St. John, "Flying Robots Are the Future of Solar," Greentech Media, November 20, 2015, https://www.greentechmedia.com/articles/read/flying-robots-the-future-solar-data-farmers-of-america.

Printed in Great Britain
by Amazon

42286584R00131